I0043914

CABI CONCISE

WASTEWATER TREATMENT

Wastewater generated from industrial processing is a major source of environmental pollution that is difficult and costly to manage. With many industries thus producing wastewater containing toxic compounds or harmful pathogens, it is mandatory to treat this wastewater to avoid contaminating sources of running clean water and impacting aquatic ecosystems. Different wastes, such as heavy metals, chemicals and organic matter, also present different challenges that require specific technological solutions.

This series examines the range of wastewater contaminants generated by various processing industries and how these can be effectively treated to ensure human health and environmental stability. The series thus provides solutions for wastewater treatment across different industries based on the most up-to-date technological and scientific advances to ensure environmental and economic sustainability.

Series Editor
Neha Srivastava,
Department of Chemical Engineering and Technology, Indian Institute of Technology (BHU), Varanasi, India
sri.neha10may@gmail.com

Magnetic Biochar for Wastewater Remediation in the Textile Industry

Dan Bahadur Pal and Ashish Kapoor

Department of Chemical Engineering, Harcourt Butler Technical University, Kanpur, Uttar Pradesh, India

CABI

CABI is a trading name of CAB International

CABI
Nosworthy Way
Wallingford
Oxfordshire OX10 8DE
UK

CABI
200 Portland Street
Boston
MA 02114
USA

Tel: +44 (0)1491 832111
E-mail: info@cabi.org
Website: www.cabi.org

T: +1 (617)682-9015
E-mail: cabi-nao@cabi.org

© CAB International 2026. All rights, including for text and data mining, AI training, and similar technologies, are reserved. No part of this publication may be reproduced in any form or by any means, electronically, mechanically, by photocopying, recording or otherwise, without the prior permission of the copyright owners.

The views expressed in this publication are those of the author(s) and do not necessarily represent those of, and should not be attributed to, CAB International (CABI). Any images, figures and tables not otherwise attributed are the author(s)' own. References to internet websites (URLs) were accurate at the time of writing.
CAB International and, where different, the copyright owner shall not be liable for technical or other errors or omissions contained herein. The information is supplied without obligation and on the understanding that any person who acts upon it, or otherwise changes their position in reliance thereon, does so entirely at their own risk. Information supplied is neither intended nor implied to be a substitute for professional advice. The reader/user accepts all risks and responsibility for losses, damages, costs and other consequences resulting directly or indirectly from using this information.

CABI's Terms and Conditions, including its full disclaimer, may be found at https://www.cabidigitallibrary.org/terms-and-conditions.

A catalogue record for this book is available from the British Library, London, UK.

Library of Congress Control Number: 2026933403

ISBN-13: 9781836993988 (hardback)
9781836993995 (paperback)
9781836994008 (ePDF)
9781836994015 (ePub)

DOI: 10.1079/9781836994015.0000

Commissioning Editor: Jamie Lee
Editorial Assistant: Theresa Regueira
Production Editor: Theresa Regueira

Typeset by Exeter Premedia Services Pvt Ltd, Chennai, India
Printed in the USA

Contents

Authors

Dr Pal is currently working as an Associate Professor, Department of Chemical Engineering, Harcourt Butler Technical University, Kanpur, Uttar Pradesh, India. He received his M. Tech in 2011 and PhD in 2017, in the field of Chemical Engineering from the Indian Institute of Technology (BHU) Varanasi, Uttar Pradesh, India. Previously, he completed his B. Tech in Chemical Engineering from UPTU, Lucknow, India, in 2006. Dr Pal completed his doctorate degree in the field of nanotechnology and catalysis. Nanofibres have very promising potential to provide benefits to nanotechnologies, energy, environment, catalysts, sensors, etc. Dr Pal and colleagues have authored 105 publications in international refereed journals, 12 books, and 55 book chapters. From all publications, they have 4240 Google citations with a 29 h-index and 63 i-10-indexes. Dr Pal's research interest is nanotechnology, catalysis, energy and environment, and waste management, with a special focus on developing process and materials using waste as raw materials. He enjoys working on bio-waste processing and value addition.

Dr Dan Bahadur Pal
B. Tech, M. Tech, PhD
Associate Professor, Department of Chemical Engineering,
Harcourt Butler Technical University, Kanpur-208002, Uttar Pradesh, India

Dr Kapoor is currently working as a Professor in the Department of Chemical Engineering at Harcourt Butler Technical University, Kanpur, Uttar Pradesh, India. He earned his B. Tech and M. Tech (dual degree programme) in Chemical Engineering from the Indian Institute of Technology (IIT) Madras, Tamil Nadu, India, in 2004. He received his PhD from the University of Illinois Urbana-Champaign (UIUC), Illinois, USA, in 2011. Before assuming his current position, Dr Kapoor worked as a Product Engineer at Lam Research Corporation, California, USA, and as a faculty member in the Department of Chemical Engineering at SRM Institute of Science and Technology, Kattankulathur, Tamil Nadu, India. With more than 12 years of experience in industry and academia, he has authored more than 75 peer-reviewed international publications and has an h-index of 18. His research interests include microfluidics, lab-on-a-chip technologies, environmental remediation, biomass valorization, and modelling and simulation.

Dr Ashish Kapoor
B. Tech, M. Tech, PhD
Professor, Department of Chemical Engineering,
Harcourt Butler Technical University, Kanpur-208002, Uttar Pradesh, India

Preface

Textile wastewater is a major environmental concern due to its high load of synthetic dyes, organic pollutants, and heavy metals, which pose significant threats to aquatic ecosystems and human health. Conventional treatment methods, such as coagulation, flocculation, and biological degradation, often suffer from inefficiencies, high operational costs, and secondary pollution. Magnetic biochar, an emerging adsorbent, offers a promising solution for textile wastewater remediation due to its enhanced adsorption capacity, reusability, and ease of separation via an external magnetic field. This book aims to provide a comprehensive exploration of the synthesis, characterization, and application of magnetic biochar in treating textile effluents.

In Chapter 1, we begin with an introduction to textile wastewater, its treatment technology and the role of magnetic biochar. Chapter 2 describes the synthesis and characterization techniques of magnetic biochar. In Chapter 3, the properties and applications of magnetic biochar in wastewater treatment are considered. Chapter 4 covers the adsorption isotherm and kinetic modelling that can be used for magnetic biochar. Finally, in Chapter 5, the mechanisms of pollutant removal using magnetic biochar are discussed.

Acknowledgements

We sincerely thank the University for supporting this book, *Magnetic Biochar for Wastewater Remediation in the Textile Industry*. The editors are grateful to the Harcourt Butler Technical University, Kanpur, India for technical facilities. We are grateful to the publishers for their support and to the academic community for providing the foundation of knowledge in this field.

Special thanks to our families, friends, and colleagues for their encouragement, and to our readers, whose curiosity inspires such efforts. We hope this book serves as a valuable resource for further wastewater treatment.

With sincere gratitude,
The Authors

Introduction to Textile Wastewater, its Treatment Technology, and the Role of Magnetic Biochar

1.1   Introduction

The textile industry is one of the oldest and most significant sectors in both the Indian and global economies. In India, textile industries hold a prominent position as one of the largest industries, contributing approximately 2.3% to the national GDP, 13% to industrial production, and more than 12% of total export earnings (Keane and Te Velde, 2008). It is the second-largest employment generator after agriculture, providing livelihoods to more than 45 million people directly and another 100 million indirectly, especially in rural and semi-urban areas (Grace Annapoorani, 2021). India is globally renowned for its strong production base in natural fibres such as cotton, jute, silk, and wool, as well as its miscellaneous handloom and handicraft traditions It is also one of the top producers of man-made fibres (MMFs) such as polyester and viscose, and has established a reputation and performs competitively in the global market (Townsend, 2020). On the global scale, the textile and apparel markets were valued at around US$1 trillion in 2024, and they continue to grow rapidly, driven by fast fashion, increasing population, and changing consumer preferences. China, India, Bangladesh, Vietnam, and Turkey are the major global textile manufacturers, with even USA and European Union (EU) being the leading consumers. The industry is crucial not only for industrialization and export-driven growth in developing economies, but also for technological innovation, particularly in the areas of technical textiles, smart fabrics, biomedical applications, and sustainable fibres (Tat *et al.*, 2022).

Despite its economic significance, the textile industry is associated with substantial environmental challenges that pose risks to human health and ecosystems. Water pollution is one of the most pervasive issues. The industry consumes massive amounts of water, particularly in dyeing and finishing processes.

Corresponding author: danbahadur.chem@gmail.com

© CAB International 2026. *Magnetic Biochar for Wastewater Remediation in the Textile Industry* (D. Bahadur Pal and A. Kapoor)

During manufacturing, large volumes of untreated or poorly treated effluents containing toxic dyes, heavy metals, salts, and surfactants are discharged into nearby water bodies (Tariq and Mushtaq, 2023). This generates groundwater contamination, destruction of aquatic life, and serious health issues such as skin disorders, gastrointestinal diseases, cancer, and hormonal imbalances among communities dependent on these water sources. Textile workers are also commonly unprotected against hazardous chemicals such as azo dyes, chlorine compounds, and phthalates, which may cause to occupational allergies, respiratory issues, reproductive toxicity, and even long-term carcinogenic effects (Burdorf *et al.*, 2006). To address these environmental and health concerns, there is need to adopt sustainable transformations in the textile sector. These include the approval of eco-friendly dyes, closed-loop water recycling systems, sustainable technologies, and a move toward organic and biodegradable fibres (Kamran *et al.*, 2025). Further, strong regulatory frameworks, environmental monitoring, and corporate responsibility initiatives are required to ensure that the economic advantages of the textile industry do not come at the cost of environmental degradation and public health. In current scenario, the conventional treatment methods such as coagulation–flocculation, activated carbon adsorption, membrane filtration, and advanced oxidation processes (AOPs) have been widely used (Sarasidis *et al.*, 2017). However, many of these technologies face problems like high operational cost, incomplete removal of contaminants, sludge generation, membrane fouling, and the need for specialized infrastructure. Additionally, the increasing presence of persistent organic pollutants (POPs), such as synthetic dyes like methylene blue and Congo red, in water bodies necessitates the development of more effective, low-cost, and environmentally friendly treatment solutions (Chen *et al.*, 2020).

Currently, biochar has emerged as a promising adsorbent for wastewater treatment due to its porous structure, high surface area, and the presence of functional groups that can interact with various contaminants. Biochar is a carbon-rich material produced by pyrolyzed biomass under limited oxygen conditions (Zaidi *et al.*, 2014). Agricultural wastes such as rice husk, banana peels, and water hyacinth have been successfully used to produce biochar. Conventional biochar such as this often shows limited adsorption capacity, limiting the extent to which pollutants can be recovered from aqueous solutions. To overcome these limitations, researchers have developed magnetic biochar or biochar modified with magnetic materials such as iron oxide (Fe_3O_4 or Fe_2O_3). Magnetic biochar not only retains the advantageous characteristics of conventional biochar but also exhibits improved surface chemistry, enhanced adsorption efficiency, and ease of separation from treated water using an external magnetic field (Wang *et al.*, 2024). Magnetic biochars have an excellent ability to adsorb dyes, heavy metals, and pharmaceuticals from polluted or contaminant water. Furthermore, they support the principles of green chemistry and the circular economy by applying agricultural or invasive biomass like water hyacinth, which is often considered an abnormality in aquatic ecosystems.

1.2 Textile Wastewater: Characteristics and Environmental Impact

1.2.1 Composition of textile wastewater

Textile wastewater is one of the challenging industrial effluents to treat due to its highly variable and contaminated composition (Ghaly *et al.*, 2014). It is generated during various stages of textile manufacturing, including dyeing, printing, finishing, and washing processes, as shown in Fig. 1.1. A major component of textile wastewater is synthetic dyes, especially azo dyes, reactive dyes, vat dyes, and disperse dyes, which are used to communicate vibrant and long-lasting colours to fabrics (Islam *et al.*, 2022). These dyes are often non-biodegradable and resist conservative biological treatment methods due to their stable molecular structures.

Azo dyes, in particular, can break down into carcinogenic aromatic amines under anaerobic conditions, posing serious risks to human health and aquatic life. In addition to dyes, textile effluents contain a range of heavy metals such as chromium, copper, zinc, lead, and cadmium, which are introduced through mordents, dye fixatives, and other chemical additives. These metals are toxic and can accumulate in living organisms, and may cause long-term ecological and health hazards. Another significant group of pollutants present in textile wastewater is surfactants, which are used as detergents and dispersants during scouring and dyeing processes. Surfactants can interpose aquatic ecosystems by reducing the surface tension of water, affecting the proper gill function of fish, and altering microbial communities. The effluent also surrounds high concentrations of salts, particularly sodium chloride and sodium sulphate, used to enhance dye uptake on fabrics (Yaseen and Scholz, 2019). Excessive salt levels increase the salinity of receiving water bodies,

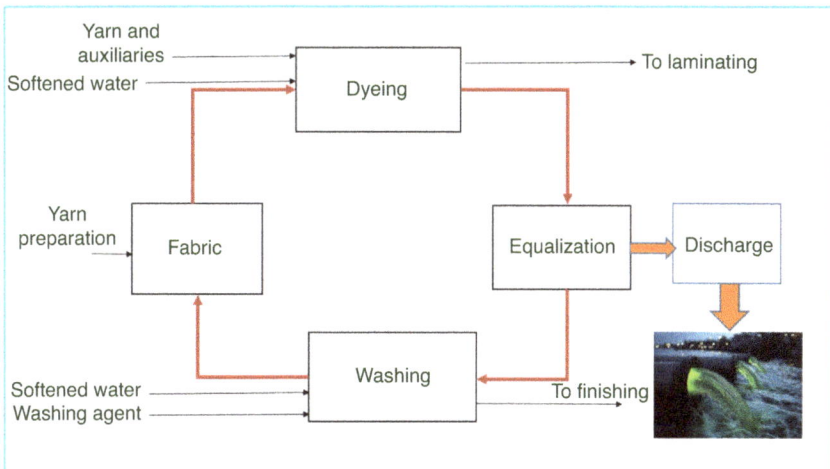

Fig. 1.1. The most commonly used manufacturing processes of the textile industry.

making them unsuitable for irrigation and aquatic life. Additionally, organic pollutants, including starch, urea, and formaldehyde-based resins, enhance chemical oxygen demand (COD) and biochemical oxygen demand (BOD) in the wastewater. All these compounds and contaminants can reduce the oxygen levels in water bodies, leading to the reduced growth of aquatic organisms. The combined presence of these pollutants' dyes, heavy metals, surfactants, salts, and organic matter makes textile wastewater highly coloured, toxic and difficult to degrade naturally (Khan *et al.*, 2023). The effective treatment of such wastewater requires integrated approaches to remove these contaminants and ensure the protection of human health.

1.2.2   Physico-chemical properties

The textile wastewater contains high levels of contaminants in the form of various types of dyes and metals that have a range of physical and chemical properties. The most important characteristic of wastewater is its high COD, which ranges from 500 to more than 2000 mg/L (Yadav *et al.*, 2013). COD represents the total quantity of oxidizable pollutants in the water, including synthetic dyes, surfactants and finishing chemicals. In parallel, textile effluents also show elevated BOD, typically greater than 200 mg/L, representing the biodegradable portion of organic contaminants. The high COD and BOD levels indicate that the wastewater is heavily polluted and can rapidly reduce the dissolved oxygen when discharged into natural water bodies, leading to aquatic life problems (Bader *et al.*, 2022). Another key parameter is total dissolved solids (TDS), which include sodium chloride, sodium sulphate, and other ionic substances used in dyeing and finishing processes. TDS levels in textile wastewater can exceed 2000 mg/L, which can adversely affect soil quality, plant growth, and freshwater ecosystems (Ali *et al.*, 2006). Additionally, pH variations are common, with values ranging widely from highly acidic (pH <4) to highly alkaline (pH >11), depending on the chemicals used in different textile treatments. These extreme pH conditions can hamper biological treatment processes, corrode infrastructure, and damage aquatic flora and fauna. The combination of high COD, BOD, TDS, and fluctuating pH makes textile wastewater toxic and non-biodegradable, requiring advanced and adaptive treatment methods that go beyond traditional physical or biological techniques.

1.2.3   Effects on ecosystems and human health

Water pollution from poorly treated textile wastewater poses severe threats to both ecosystems and human health. One of the most immediate effects is the toxicity to aquatic life, as textile effluents often contain synthetic dyes, heavy metals, surfactants, and other harmful organic chemicals. These substances can reduce the penetration of sunlight into water bodies, reducing photosynthesis in aquatic plants and leading to a decline in dissolved oxygen levels (Cooper, 1993). Fish and other aquatic organisms suffer from respiratory distress, reproductive failure, and mortality due to the toxic effects of heavy metals like chromium, as well as persistent organic pollutants. Over time, some of these substances amass in the tissues of aquatic organisms, a process known as bioaccumulation, and can amplify up the food chain, eventually impacting birds, animals, and even humans who consume contaminated water or seafood. This can increase long-term health risks such as cancer, neurological disorders, hormonal imbalances, and organ damage. In communities located near textile industrial zones, exposure to polluted water can result in skin irritation, gastrointestinal problems, and chronic diseases. Mostly, the high levels of COD and TDS in the water degrade overall water quality. The cumulative effects of such pollution not only have biodiversity and ecosystem consequences but also compromise public health and sustainable development.

1.3 Current Regulations and Treatment Standards of Textile Wastewater

The treatment and discharge of textile wastewater are regulated by national and international environmental agencies to minimize the impact on water bodies and public health. In India, the Central Pollution Control Board (CPCB) sets the effluent discharge standards under the Environment (Protection) Rules, 1986, specifically targeting the textile sector. Key parameters and their permissible limits include pH of 5.5–9.0, $BOD \leq 30\,mg/L$, $COD \leq 250\,mg/L$, $TSS \leq 100\,mg/L$, oil and grease $\leq 10\,mg/L$, and TDS $\leq 2100\,mg/L$ (Daud *et al.*, 2015). For dye units, colour removal is also crucial, though specific numerical standards for colour are less commonly enforced and vary by state. Globally, the EU follows stringent standards under the Urban Waste Water Treatment Directive (91/271/EEC) and REACH Regulation (Registration, Evaluation, Authorisation, and Restriction of Chemicals), which restrict the use of hazardous substances, such as certain azo dyes. In the USA, the Environmental Protection Agency (EPA) regulates textile effluent under the Clean Water Act (CWA) through the Effluent Guidelines and Standards for the Textile Mills. The EPA mandates limits on BOD, COD, TSS, pH, and heavy metals for direct and indirect dischargers to surface waters or municipal treatment plants. In recent years, growing concerns over environmental sustainability have led to additional pressure from voluntary international standards and certifications. These include OEKO-TEX®, ZDHC (Zero Discharge of Hazardous Chemicals), and bluesign®, which promote safer chemical management and water treatment in textile production. Many exporting companies now adopt best available techniques (BATs) to meet not only national compliance but also global buyer requirements, especially in the EU and North America. Despite these regulations, enforcement and compliance often remain weak, especially in developing countries. Many small- and medium-scale textile units lack adequate effluent treatment plants (ETPs) or fail to operate them efficiently. Hence, stricter monitoring, adoption of advanced treatment technologies like membrane filtration, adsorption using biochar, and increased awareness are essential to ensure safe and sustainable textile wastewater management.

1.4   Conventional Textile Wastewater Treatment Methods

1.4.1   Physical methods for textile wastewater treatment

Physical treatment methods, such as filtration and sedimentation, are the initial steps in textile wastewater treatment and are primarily used to remove suspended solids and larger particulate matter before further chemical or biological processes. These methods are simple, cost-effective, and crucial for reducing the load on downstream treatment units. Sedimentation involves the gravitational settling of heavier particles from the wastewater in a settling tank or sedimentation basin. In textile industries, sedimentation is typically used to remove fibres, lint, and other coarse solids that are introduced during the washing and processing of fabrics (Shammas et al., 2005). The wastewater is allowed to stand still or flow slowly through a tank, where solid particles gradually settle at the bottom, forming a sludge that can be periodically removed and treated, or disposed. The clear water at the top is then directed to further treatment stages. Sedimentation is often enhanced by coagulation and flocculation, where chemical agents like alum or poly aluminium chloride are added to agglomerate fine particles into larger floc or molecules that settle more easily.

Filtration is used to physically screen out smaller suspended particles that do not settle by gravity. Various types of filters are employed, such as sand filters, multimedia filters, activated carbon filters, and membrane filters (Rahman et al., 2022). Sand and multimedia filters are commonly used to remove turbidity and residual solids, while activated carbon filters also help in adsorbing some organic compounds and colorants. Filtration not only improves the aesthetic quality of the treated water but also prevents clogging and fouling in advanced treatment systems like reverse osmosis or adsorption units. Although physical methods alone are not sufficient to remove dissolved pollutants like dyes, salts, or heavy metals, they serve an essential pre-treatment role, as shown in Fig. 1.2. By significantly reducing the solid load, they improve the efficiency and longevity of subsequent treatment processes such as chemical oxidation, biological treatment, or adsorption.

1.4.2   Chemical methods for textile wastewater treatment

There are some chemical methods, such as coagulation/flocculation and AOPs that are most commonly used to treat textile wastewater, especially the removal of dissolved pollutants, dyes, and other available contaminants that physical methods like sedimentation and filtration cannot eliminate efficiently (Carneiro et al., 2010). Coagulation and flocculation are commonly employed in textile wastewater treatment to remove suspended solids, colloidal particles, and certain dissolved substances, including organic pollutants and dyes. In the

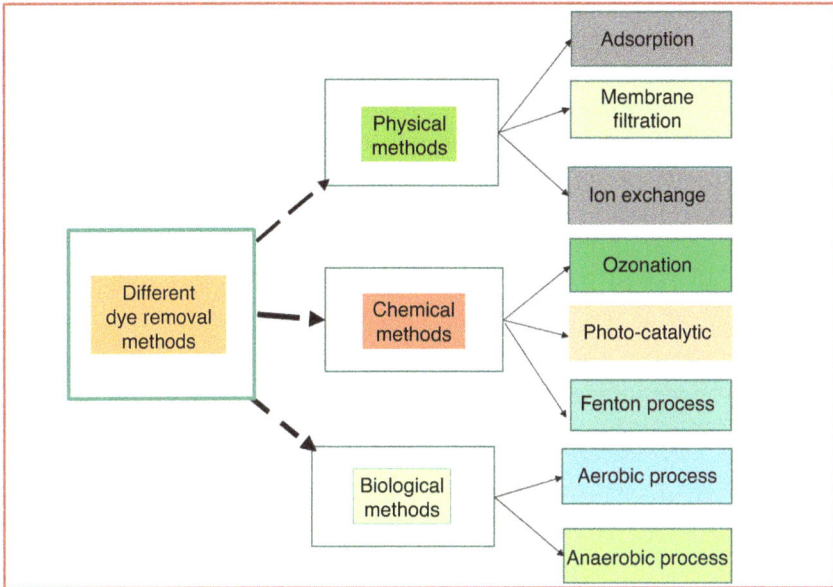

Fig. 1.2. Conventional methods for treating textile water.

first of the two steps, coagulation, a coagulant, typically an inorganic metal salt like aluminium sulphate or ferric chloride, is added to the wastewater (Bakar and Halim, 2013). These coagulants neutralize the charges on suspended particles, causing them to aggregate or clump together. After coagulation, the flocculants, which are usually polymers such as polyacrylamide or other organic compounds, are added. Flocculants help to bind the smaller coagulated particles into larger floc-like aggregates that can settle or be removed by filtration. This step enhances the removal of colour, suspended solids, and other contaminants in textile wastewater. The chemical coagulation/flocculation process is particularly effective for treating dyes and organic pollutants, as well as reducing the COD and BOD of the effluent (Abdalla and Hammam, 2014). However, it may not be sufficient to remove dissolved heavy metals or more resistant pollutants, which require combined treatment methods.

1.4.2.1   Advanced oxidation processes (AOPs)

AOPs are a group of chemical treatment technologies that produce highly reactive hydroxyl radicals (•OH) to break down organic pollutants in wastewater. These processes are particularly effective for the removal of dyes, pesticides, pharmaceuticals, and other organic contaminants that do not easily degrade through biological or chemical methods. Some AOPs used in textile wastewater treatment include ozone, which is a powerful oxidant that, when introduced into wastewater, breaks down organic contaminants through direct oxidation or the generation of hydroxyl radicals (Wang and Xu, 2012). Ozone is particularly effective in decolorizing textile effluents and can break down important organic dyes, such as azo dyes, into simpler or less toxic compounds. Fenton's reagent is a mixture of hydrogen peroxide (H_2O_2) and ferrous sulphate (Fe^{2+}), which generates hydroxyl radicals in the presence of iron ions (Ahmad and Bensalah, 2022). This process is highly effective in oxidizing organic pollutants, including azo dyes and other aromatic compounds commonly used in textile wastewater. In photo-catalytic processes, a semiconductor catalyst, typically titanium dye oxide TiO_2, is activated by ultraviolet (UV) light to generate hydroxyl radicals. These radicals attack and degrade organic pollutants in textile wastewater which make photo-catalysis an effective method for dye degradation and mineralization. This process is particularly useful for degrading colorants and organic matter in textile effluents. AOPs are particularly advantageous for treating persistent dyes and non-biodegradable organic compounds that are difficult to remove through conventional methods. However, these processes are energy-intensive and can be costly, limiting their widespread use in the textile industry unless integrated with other, more cost-effective treatment steps. Both coagulation/flocculation and advanced oxidation processes are effective chemical methods for textile wastewater treatment, each serving specific roles in addressing the diverse contaminants found in textile effluents. Coagulation/flocculation is effective for removing suspended solids and certain dissolved pollutants, while AOPs offer a more advanced solution for breaking down intractable organic pollutants and dyes (Karam *et al.*, 2021). These chemical methods, when combined with physical and biological treatment processes, form a wide-ranging approach to managing the complex nature of textile wastewater and moderating its environmental impact.

1.4.3   Biological treatment methods for textile wastewater

Biological treatment methods are an important group of methods for the sustainable treatment of textile wastewater because they utilize microorganisms to break down organic pollutants, particularly those that are biodegradable, like dyes, detergents, and other organic compounds. Some biological methods that are used to treat textile wastewater are described here.

1.4.3.1   Activated sludge process

The activated sludge process (ASP) is one of the most widely used biological treatment methods for wastewater, including textile effluents. In this process, wastewater is aerated in a tank to dissolve oxygen for microbial growth, typically bacteria and protozoa, which break down the organic pollutants (Singh *et al.*, 2021). The process involves the following key steps. First, aeration occurs, where the wastewater is mixed with air (oxygen) to support the growth of microorganisms to convert organic pollutants into biomass and simpler compounds. In the sedimentation step, the mixture is allowed to settle in a secondary clarifier, where the biomass (sludge) separates from the treated water. The clear water is then decanted, while the sludge is either recirculated to the aeration tank or removed for further treatment. The ASP is effective for removing organic pollutants, suspended solids, and some types of dyes, particularly those that are biodegradable (Hanafi and Sapawe, 2020). However, it may not be sufficient for removing persistent and non-biodegradable compounds commonly found in textile wastewater, like certain synthetic dyes and chemicals. Therefore, it is often used in conjunction with other treatment methods like coagulation/flocculation or advanced oxidation.

1.4.3.2   Aerobic/anaerobic treatments

Aerobic and anaerobic treatments are biological processes that rely on microorganisms to break down pollutants in the presence or absence of oxygen. These treatments are particularly effective for reducing the COD and BOD of textile wastewater (Fig. 1.3).

In the aerobic treatment, the process occurs in the presence of oxygen and is typically used for the degradation of organic pollutants. Aerobic bacteria are cultured in reactors like sequencing batch reactors (SBRs) or activated sludge systems, where they consume the organic matter in the wastewater. This method is effective for the removal of biodegradable organic compounds and is commonly used in municipal and industrial wastewater treatment systems. The primary advantage of aerobic treatment is its speed and effectiveness in reducing organic loads.

The anaerobic treatment occurs in the absence of oxygen. In this process, anaerobic bacteria break down organic materials, producing biogas (methane) as a by-product. Anaerobic treatment is highly effective for high-strength wastewater with high COD levels, such as textile effluents, as it requires less energy than aerobic treatment. Anaerobic systems are particularly useful for removing organic pollutants in textile wastewater, including surfactants and some types of dyes that are not easily treated by aerobic processes (Rashid

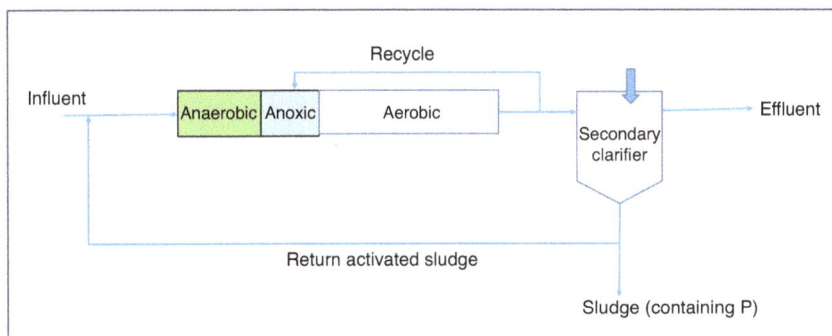

Fig. 1.3. A schematic diagram of an aerobic and anaerobic treatment plant.

et al., 2020). Anaerobic processes are typically used in UASB (up flow anaerobic sludge blanket) reactors, anaerobic filters, or anaerobic lagoons. One of the key benefits of anaerobic treatment is that it produces methane gas, which can be captured and used as an energy source, making the process more cost-effective. However, anaerobic treatment is typically slower and requires careful control of environmental conditions such as temperature, pH, and hydraulic retention time. Both the ASP and aerobic/anaerobic treatments are effective biological methods for treating textile wastewater (Sarayu and Sandhya, 2012). The ASP is commonly used to degrade organic matter and reduce BOD and COD levels. However, for more resistant pollutants, especially non-biodegradable compounds, a combination of aerobic and anaerobic treatments can offer a more comprehensive solution. While aerobic processes are faster and efficient, anaerobic treatments provide energy savings and are effective for high-strength wastewater. In practice, a combination of these biological methods, often integrated with physical and chemical treatments, is essential for achieving high-quality effluent that meets environmental standards. Table 1.1 shows the advantages and disadvantages of physical, chemical, and biological treatments (Grant *et al.*, 1930; Daskalopoulos *et al.*, 1997; Bellanthudawa *et al.*, 2023; Reddy *et al.*, 2024).

1.5   Need for and Role of Magnetic Biochar

Biochar has gained increasing attention as an effective and sustainable material for wastewater treatment, particularly in the textile industry where conventional methods often fall short (Kumar *et al.*, 2025). Textile effluents are complex and contain persistent pollutants such as synthetic dyes, heavy metals, and various organic compounds that are not easily removed through traditional treatment methods like coagulation, filtration, or biological processes. These conventional methods often involve high operational costs, produce large amounts of sludge, and fail to completely remove non-biodegradable contaminants. In this context, magnetic biochar offers important alternative owing to its high surface area, porosity, and the presence of functional groups that increased adsorption capacity (Table 1.1) (Katibi *et al.*, 2024). Produced by pyrolysing biomass and incorporating magnetic particles like iron oxide (Fe_3O_4), magnetic biochar not only adsorbs a wide range of pollutants but also allows for easy recovery from treated water using a simple magnetic field, thus reducing the risk of secondary pollution. Furthermore, it plays a multifunctional role by adsorbing dyes and heavy metals, acting as a catalyst in AOPs, buffering pH levels, and minimizing sludge production. This process is low cost and eco-friendly nature, combined with having excellent pollutant removal performance and reusability, making magnetic biochar a valuable and scalable solution for addressing the growing environmental concerns posed by textile wastewater.

Table 1.1. Advantages and disadvantages of physical, chemical, and biological treatments (Grant *et al.*, 1930; Daskalopoulos *et al.*, 1997; Bellanthudawa *et al.*, 2023; Reddy *et al.*, 2024).

Treatment method	Limitations of conventional method	Benefits of magnetic biochar
Physical (filtration, sedimentation)	Limited removal of dissolved dyes and metals	High adsorption of dissolved dyes and metals
	Clogging of filters	Removes micro-pollutants
	Low efficiency for micro-pollutants	Minimal clogging
Chemical (e.g. coagulation/ flocculation, AOPs)	High chemical usage	Reduces chemical dependency
	Excess sludge generation	Low sludge production
	Costly disposal	Acts as a catalyst in AOPs
Biological (e.g. activated sludge, aerobic/anaerobic)	Ineffective for non-biodegradable compounds	Adsorbs non-biodegradable dyes and metals
	Slow processing time	Faster action
	Sensitive to toxic substances	Stable performance under toxicity
Cross-cutting advantages		Easily recoverable using a magnet
		Reusable over multiple cycles
		Environmentally friendly

1.6   Conclusion

The textile industry, while playing a vital role in economic development and employment, especially in nations like India, significantly contributes to environmental degradation through the release of untreated and complex wastewater. Characterized by high concentrations of dyes, heavy metals, salts, surfactants, and fluctuating physicochemical parameters, textile effluents pose serious ecological and public health risks. Traditional physical, chemical, and biological treatment methods, though widely practiced, often suffer from inefficiencies such as high operational costs, excessive sludge generation, and limited pollutant removal. In response to these challenges, biochar and magnetic biochar have emerged as promising, sustainable alternatives. Derived from biomass through pyrolysis, these materials offer high adsorption capacity, cost-effectiveness, and environmental compatibility. The magnetic modification further enhances their practicality by enabling easy recovery and reuse. There is a need to move beyond conventional methods and adopt advanced, eco-friendly technologies like magnetic biochar to address the pressing issue of textile wastewater pollution. Integrating such sustainable materials into wastewater management frameworks not only mitigates environmental and health hazards but also supports long-term industrial sustainability and circular economy goals.

References

Abdalla, K.Z. and Hammam, G. (2014) Correlation between biochemical oxygen demand and chemical oxygen demand for various wastewater treatment plants in Egypt to obtain the biodegradability indices. *International Journal of Sciences: Basic and Applied Research* 13(1), 42–48.

Ahmad, M.I. and Bensalah, N. (2022) Insights into the generation of hydroxyl radicals from H2O2 decomposition by the combination of Fe2+ and chloranilic acid. *International Journal of Environmental Science and Technology* 19(10), 10119–10130. DOI: 10.1007/s13762-021-03822-0.

Ali, S., Nadeem, R., Bhatti, H.N., Hayat, S.H.A.U.K.A.T., Ali, S. *et al.* (2006) Analyses and treatment of textile effluents. *International Journal of Agriculture and Biology* 8(5), 1–4.

Bader, A.C., Hussein, H.J. and Jabar, M.T. (2022) BOD: COD ratio as indicator for wastewater and industrial water pollution. *International Journal of Special Education* 37(3), 2164–2171.

Bakar, A.F.A. and Halim, A.A. (2013) Treatment of automotive wastewater by coagulation-flocculation using poly-aluminum chloride (PAC), ferric chloride (FeCl 3) and aluminum sulfate (alum). In: *AIP Conference Proceedings*, American Institute of Physics, pp. 524–529 (Vol. 1571, No. 1).

Bellanthudawa, B.K.A., Nawalage, N.M.S.K., Handapangoda, H.M.A.K., Suvendran, S., Wijayasenarathne, K.A.S.H. *et al.* (2023) A perspective on biodegradable and non-biodegradable nanoparticles in industrial sectors: Applications, challenges, and future prospects. *Nanotechnology for Environmental Engineering* 8(4), 975–1013. DOI: 10.1007/s41204-023-00344-7.

Burdorf, A., Figa-Talamanca, I., Jensen, T.K. and Thulstrup, A.M. (2006) Effects of occupational exposure on the reproductive system: Core evidence and practical implications. *Occupational Medicine* 56(8), 516–520. DOI: 10.1093/occmed/kql113.

Carneiro, P.A., Umbuzeiro, G.A., Oliveira, D.P. and Zanoni, M.V.B. (2010) Assessment of water contamination caused by a mutagenic textile effluent/dyehouse effluent bearing disperse dyes. *Journal of Hazardous Materials* 174(1–3), 694–699. DOI: 10.1016/j.jhazmat.2009.09.106.

Chen, T., Zhang, J., Wang, Z., Zhao, R., He, J. *et al.* (2020) Oxygen-enriched gasification of lignocellulosic biomass: Syngas analysis, physicochemical characteristics of the carbon-rich material and its utilization as an anode in *Lithium ion* battery. *Energy* 212, 118771. DOI: 10.1016/j.energy.2020.118771.

Cooper, C.M. (1993) Biological effects of agriculturally derived surface water pollutants on aquatic systems—a review. *Journal of Environmental Quality* 22(3), 402–408. DOI: 10.2134/jeq1993.00472425002200030003x.

Daskalopoulos, E., Badr, O. and Probert, S.D. (1997) Economic and environmental evaluations of waste treatment and disposal technologies for municipal solid waste. *Applied Energy* 58(4), 209–255. DOI: 10.1016/S0306-2619(97)00053-6.

Daud, Z., Awang, H., Nasir, N., Ridzuan, M.B. and Ahmad, Z. (2015) Suspended solid, color, COD and oil and grease removal from biodiesel wastewater by coagulation and flocculation processes. *Procedia-Social and Behavioral Sciences* 195, 2407–2411.

Ghaly, A.E., Ananthashankar, R., Alhattab, M.V.V.R. and Ramakrishnan, V.V. (2014) Production, characterization and treatment of textile effluents: A critical review. *Journal of Chemical Engineering & Process Technology* 5(1), 1–19. DOI: 10.4172/2157-7048.1000182.

Grace Annapoorani, S. (2021) Sustainable development in the handloom industry. In: *Handloom Sustainability and Culture: Artisanship and Value Addition.* Springer Singapore, Singapore, pp. 95–118.

Grant, S., Hurwitz, E. and Mohlman, F.W. (1930) The oxygen requirements of the activated sludge process. *Sewage Works Journal* 2(2), 228–244.

Hanafi, M.F. and Sapawe, N. (2020) A review on the current techniques and technologies of organic pollutants removal from water/wastewater. *Materials Today: Proceedings* 31, A158–A165. DOI: 10.1016/j.matpr.2021.01.265.

Islam, M.T., Islam, T., Islam, T. and Repon, M.R. (2022) Synthetic dyes for textile colouration: Process, factors and environmental impact. *Textile & Leather Review* 5, 327–373. DOI: 10.31881/TLR.2022.27.

Kamran, F., Afshar, H. and Shahi, F. (2025) *Recent advances and applications of sustainable and recyclable polymers.* Polymer Engineering & Science.

Karam, A., Bakhoum, E.S. and Zaher, K. (2021) Coagulation/flocculation process for textile mill effluent treatment: Experimental and numerical perspectives. *International Journal of Sustainable Engineering* 14(5), 983–995. DOI: 10.1080/19397038.2020.1842547.

Katibi, K.K., Shitu, I.G., Yunos, K.F.M., Azis, R.S., Iwar, R.T. *et al.* (2024) Unlocking the potential of magnetic biochar in wastewater purification: A review on the removal of bisphenol a from aqueous solution. *Environmental Monitoring and Assessment* 196(5), 492. DOI: 10.1007/s10661-024-12574-6.

Keane, J. and Te Velde, D.W. (2008) The role of textile and clothing industries in growth and development strategies. *Overseas Development Institute* 7(1), 141–147.

Khan, W.U., Ahmed, S., Dhoble, Y. and Madhav, S. (2023) A critical review of hazardous waste generation from textile industries and associated ecological impacts. *Journal of the Indian Chemical Society* 100(1), 100829. DOI: 10.1016/j.jics.2022.100829.

Kumar, A., Kapoor, A., Kumar Rathoure, A., Lal Devnani, G. and Pal, B.D. (2025) Organic compounds removal using magnetic biochar from textile industries based wastewater–a comprehensive review. *Sustainable Processes Connect* 1, 0012. DOI: 10.69709/SusProc.2025.147565.

Rahman, A., Salman, A., Nainggolan, R., Siregar, S.A., Wan Ismail, W.Z. *et al.* (2022) Water treatment process using manganese zeolite filter, activated carbon filter, and silica sand filter. *International Journal of Technical Vocational and Engineering Technology* 3(3), 1–7.

Rashid, T.U., Kabir, S.M.F., Biswas, M.C. and Bhuiyan, M.A.R. (2020) Sustainable wastewater treatment via dye–surfactant interaction: A critical review. *Industrial & Engineering Chemistry Research* 59(21), 9719–9745. DOI: 10.1021/acs.iecr.0c00676.

Reddy, C.N., Surabhi, D., Charan, M.C.K., Balla, R.P., Katla, H. *et al.* (2024) Aerobic and anaerobic digestion of textile industry wastewater. In: *Aerobic and Anaerobic Microbial Treatment of Industrial Wastewater.* CRC Press, pp. 93–136.

Sarasidis, V.C., Plakas, K.V. and Karabelas, A.J. (2017) Novel water-purification hybrid processes involving *in-situ* regenerated activated carbon, membrane separation and advanced oxidation. *Chemical Engineering Journal* 328, 1153–1163. DOI: 10.1016/j.cej.2017.07.084.

Sarayu, K. and Sandhya, S.J.A.B. (2012) Current technologies for biological treatment of textile wastewater--a review. *Applied Biochemistry and Biotechnology* 167(3), 645–661. DOI: 10.1007/s12010-012-9716-6.

Shammas, N.K., Kumar, I.J., Chang, S.Y. and Hung, Y.T. (2005) Sedimentation. In: *Physicochemical Treatment Processes*. Humana Press, Totowa, NJ, pp. 379–429.

Singh, G., Singh, A., Singh, P., Shukla, R., Tripathi, S. *et al.* (2021) The fate of organic pollutants and their microbial degradation in water bodies. *Pollutants and Water Management: Resources, Strategies and Scarcity* 210–240. DOI: 10.1002/9781119693635.ch9.

Tariq, A. and Mushtaq, A. (2023) Untreated wastewater reasons and causes: A review of most affected areas and cities. *International Journal of Chemical and Biochemical Sciences* 23(1), 121–143.

Tat, T., Chen, G., Zhao, X., Zhou, Y., Xu, J. *et al.* (2022) Smart textiles for healthcare and sustainability. *ACS Nano* 16(9), 13301–13313. DOI: 10.1021/acsnano.2c06287.

Townsend, T. (2020) World natural fibre production and employment. In: *Handbook of Natural Fibres*. Woodhead Publishing, pp. 15–36.

Wang, J.L. and Xu, L.J. (2012) Advanced oxidation processes for wastewater treatment: Formation of hydroxyl radical and application. *Critical Reviews in Environmental Science and Technology* 42(3), 251–325. DOI: 10.1080/10643389.2010.507698.

Wang, B., Ma, Y., Cao, P., Tang, X. and Xin, J. (2024) Ball milling and magnetic modification boosted methylene blue removal by Biochar obtained from water hyacinth: Efficiency, mechanism, and application. *Molecules* 29(21), 5141. DOI: 10.3390/molecules29215141.

Yadav, A., Mukherji, S. and Garg, A. (2013) Removal of chemical oxygen demand and color from simulated textile wastewater using a combination of chemical/physicochemical processes. *Industrial & Engineering Chemistry Research* 52(30), 10063–10071. DOI: 10.1021/ie400855b.

Yaseen, D.A. and Scholz, M. (2019) Textile dye wastewater characteristics and constituents of synthetic effluents: A critical review. *International Journal of Environmental Science and Technology* 16(2), 1193–1226. DOI: 10.1007/s13762-018-2130-z.

Zaidi, N.S., Sohaili, J., Muda, K. and Sillanpää, M. (2014) Magnetic field application and its potential in water and wastewater treatment systems. *Separation & Purification Reviews* 43(3), 206–240. DOI: 10.1080/15422119.2013.794148.

Synthesis and Characterization Techniques of Magnetic Biochar

2

2.1 Introduction

Magnetic biochar is a novel, multifunctional material that is used by modifying conventional biochar with magnetic properties, typically with the help of iron-based nanoparticles such as magnetite (Fe_3O_4) (Liu *et al.*, 2023a). This modification increases the importance of biochar, especially in environmental and industrial applications like wastewater treatment, soil remediation, and catalysis. The conventional biochar, though useful for its high surface area, porosity, and carbon content, has major limitations due to its poor recoverability after application in liquid-phase systems, often needing labour-intensive filtration or centrifugation (Hamamah and Grützner, 2022). Magnetic biochar, however, can be effectively and rapidly separated from used solutions using a simple external magnetic field, making it effective and cost-effective. The synthesis of magnetic biochar is usually achieved through techniques such as co-precipitation, impregnation followed by pyrolysis, or hydrothermal methods, where iron salts such as $FeCl_3$ and $FeSO_4$ are incorporated into the biomass or biochar and then subjected to thermal or chemical treatment to form magnetically active particles (Tripathi *et al.*, 2016), as shown in Fig. 2.1. The primary desire for magnetizing biochar lies in enhancing its separation ease and recyclability, while at the same time improving its adsorption power for pollutants. Magnetic biochar has excellent reusability and can withstand multiple adsorption–desorption cycles without substantial deterioration in performance, which promotes sustainability and economic feasibility. The presence of magnetic particles often alters the surface chemistry and morphology of biochar, introducing functional groups and active sites that increase its affinity for a wide range of contaminants including heavy metals like cadmium (Huang *et al.*, 2019). These improvements are also recognized to increase surface area, enhance porosity, and improve dispersibility in aqueous media, which collectively lead to faster and more effective adsorption.

Corresponding author: danbahadur.chem@gmail.com

© CAB International 2026. *Magnetic Biochar for Wastewater Remediation in the Textile Industry* (D. Bahadur Pal and A. Kapoor)

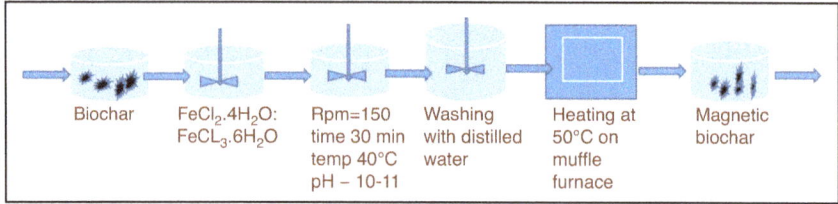

Fig. 2.1. Schematic diagram of the preparation of magnetic biochar.

Magnetic biochar not only acts as an adsorbent but also serves as a catalyst or catalyst support in advanced oxidation processes such as Fenton-like reactions, which are important for breaking down obstinate organic pollutants in industrial effluents (Hussain *et al.*, 2021). Furthermore, the material can be tailored by varying parameters like the concentration of iron salts, pH, temperature, and pyrolysis conditions, allowing for application-specific optimization. Its usefulness extends to agricultural uses such as nutrient preservation in soil and even in energy systems as an electrode material. Importantly, magnetic biochar contributes to the goals of sustainable and green chemistry because it permits resource recovery, pollutant removal, and waste biomass valorization – all in a reusable and recyclable framework (de Souza Mesquita *et al.*, 2024).

The properties of magnetic biochar are typically determined through characterization techniques such as X-ray diffraction (XRD) to validate the crystalline phases of iron oxides, scanning electron microscopy (SEM) to observe surface morphology, energy-dispersive X-ray spectroscopy (EDX) for elemental composition, Fourier-transform infrared spectroscopy (FTIR) for surface functional groups, Brunauer–Emmett–Teller (BET) analysis for surface area, and vibrating sample magnetometer (VSM) to estimate the magnetic behaviour. Overall, the development of magnetic biochar represents a significant advancement in the field of environmental engineering and materials science, offering a simple, effective solution for pollution control, especially in regions where low-cost, scalable, and eco-friendly materials are in high demand (Rath *et al.*, 2021). By converting agricultural or aquatic biomass such as rice husks, coconut shells, or water hyacinth into magnetic biochar, waste streams are turned into valuable remediation tools, reinforcing circular economy principles and addressing critical environmental challenges.

2.2   Common Synthesis Methods

2.2.1   Co-precipitation method

The co-precipitation method is one of the most used and effective techniques for synthesizing magnetic biochar, primarily due to its sustainability, cost-effectiveness, and ability to uniformly deposit magnetic nanoparticles like Fe_3O_4 (magnetite) onto the surface of biochar. This process involves mixing pre-prepared biochar with a solution containing iron salts, typically ferrous (Fe^{2+}) and ferric (Fe^{3+}) ions in a molar ratio of 1:2 (often using salts like $FeSO_4 \cdot 7H_2O$ and $FeCl_3 \cdot 6H_2O$), and then introducing a base such as sodium hydroxide (NaOH) or ammonium hydroxide (NH_4OH) to precipitate magnetite nanoparticles directly onto the surface or within the porous structure of the biochar (Lan *et al.*, 2022). The basic chemical reaction that forms Fe_3O_4 under alkaline conditions is (Eqn 1):

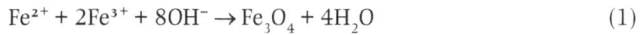

$$Fe^{2+} + 2Fe^{3+} + 8OH^- \rightarrow Fe_3O_4 + 4H_2O \qquad (1)$$

This process is typically under controlled conditions of pH, temperature, and reaction time to ensure effective precipitation and proper magnetic nano-particle formation. The pH is critical and is usually maintained between 9 and 11 because a highly alkaline medium promotes the complete precipitation of magnetite and prevents the formation of unwanted by-products like $Fe(OH)_3$ (Ziemniak *et al.*, 1995). The reaction temperature is between 60°C and 90°C, with moderate heating facilitating uniform nucleation and growth of the Fe_3O_4 particles. Reaction time can vary from 30 min to several hours, depending on the desired particle size and distribution, as well as the porosity and surface chemistry of the biochar. During the reaction, the iron ions interact with the functional groups on the biochar surface such as hydroxyl ($-OH$), carboxyl ($-COOH$), and carbonyl ($=O$) which helps in anchoring the magnetic particles and improves the dispersion of Fe_3O_4, thereby increasing the active surface area for adsorption (Awang *et al.*, 2023). Once the reaction is complete, the mixture is cooled, and the magnetic biochar is separated using a strong magnet, washed repeatedly with deionized water and ethanol to remove residual salts and unreacted chemicals, and finally dried (either at room temperature or under mild heating) for storage and application. The co-precipitation method not only ensures good magnetic saturation of the biochar but also improves its adsorption capacity and surface reactivity, making it highly efficient in removing contaminants like heavy metals, dyes, and organic pollutants from aqueous media (Zhang *et al.*, 2023). The deposited magnetite also imparts recyclability and reusability, allowing the biochar to be recovered and reused multiple times with minimal loss of performance. Therefore, controlling the reaction conditions – particularly iron salt concentration, pH, temperature, and contact time – is crucial to tailor the physicochemical and magnetic properties of the resulting magnetic biochar for specific environmental or industrial applications.

2.2.2   Impregnation and pyrolysis method

Synthesizing magnetic biochar involves soaking raw biomass or pre-formed biochar in iron salt solutions such as $FeCl_3$ or $Fe(NO_3)_3$ to allow iron ions to infiltrate the material's porous structure, followed by drying and pyrolyzing the impregnated biomass at controlled temperatures (typically 400–700°C) in an inert atmosphere (Hunter, 2023). During pyrolysis, the organic matter converts to carbon-rich biochar while the iron salts thermally decompose to form magnetic nanoparticles (Fe_3O_4 or γ-Fe_2O_3), which become strongly embedded in the biochar matrix. This method offers uniform distribution of magnetic particles, high thermal stability, and improved adsorption performance, making the resulting magnetic biochar ideal for applications in wastewater treatment, heavy metal and dye removal, and catalysis. This process is simple, scalable, and removes the need for post-synthesis magnetization, which makes it a preferred technique for preparing durable, recyclable magnetic biochar.

To make the process more cost-effective, mixing biomass with iron salts before pyrolysis is necessary. The one-step pyrolysis with iron precursor method is a simplified and cost-effective approach for synthesizing magnetic biochar, where biomass is directly mixed with iron salts such as ferric chloride ($FeCl_3$) and ferric nitrate ($Fe(NO_3)_3$) before pyrolysis, eliminating the need for separate impregnation and thermal treatment steps. In this method, the biomass (e.g. agricultural waste, aquatic plants like water hyacinth) is blended uniformly with a solution or slurry of iron salts and then dried to remove excess moisture (Guna *et al.*, 2017). The mixture is subsequently pyrolyzed in an inert or low-oxygen environment (e.g. nitrogen) at temperatures typically ranging from 400°C to 700°C. During the thermal process, the iron precursors decompose and react with the evolving gases and biochar matrix to form magnetic iron oxides (mainly Fe_3O_4 or γ-Fe_2O_4) *in situ*, which become anchored onto or embedded within the biochar structure. This single-step method is advantageous due to its operational simplicity, reduced chemical and energy input, and strong integration of magnetic particles, making the final product effective for environmental applications such as adsorption of pollutants, catalysis, and water purification. It is particularly suitable for large-scale or low-cost production scenarios where efficiency and minimal processing are prioritized (Bramsiepe *et al.*, 2012).

2.2.3 Hydrothermal synthesis

Hydrothermal synthesis is a versatile and efficient method for producing magnetic biochar by combining biochar (or raw biomass) with magnetic precursors (such as $FeCl_3$, $FeSO_4$, or $Fe(NO_3)_3$ in an aqueous medium inside a sealed autoclave or hydrothermal reactor, which is then heated to moderate temperatures (typically 120–250°C) under self-generated pressure (Almeida, 2010). This process represents the natural geological formation conditions that enhance controlled crystal growth and uniform deposition of magnetic nanoparticles (Fe_3O_4) onto the surface or within the pores of biochar. During the hydrothermal reaction, iron ions undergo hydrolysis and nucleation, making well-dispersed magnetic crystals that are integrated tightly into the carbon matrix (Liu *et al.*, 2015). The relatively low temperature and aqueous environment reduce particle aggregation and allow for the formation of nanostructured iron oxides with high surface reactivity. Hydrothermal synthesis is particularly valuable for modifying the size, morphology, and crystallinity of the magnetic particles and can be modified by adjusting parameters like pH, reaction time, precursor concentration, and temperature (Torres-Gómez *et al.*, 2019) This method yields magnetic biochar with excellent surface chemistry, high magnetization, and strong pollutant affinity, making it ideal for adsorption of dyes and heavy metals, catalytic degradation, and sensor applications.

2.2.4   Sol-gel methods

This process involves the hydrolysis and polycondensation of metal lakesides or metal salts (e.g. $FeCl_3$, $Fe(NO_3)_3$, or iron lakesides) in a liquid 'sol' phase, which gradually transforms into a solid 'gel' network that coats the biochar surface. The gel, containing precursors for magnetic oxides, is then dried and thermally treated (calcined) to form nanostructured iron oxide coatings on the biochar matrix (Weidner *et al.*, 2022). During this transformation, the sol-gel technique allows for precise control over particle size, morphology, and coating thickness. The iron oxide nanoparticles (usually Fe_2O_3 or Fe_3O_4) produced by sol-gel coating are often highly dispersed, creating a fine, conformal magnetic layer on the porous biochar. This method is particularly useful when ultra-fine particle distribution and surface uniformity are critical, such as in high-performance catalysis, sensor development, or selective adsorption applications (Cho *et al.*, 2016). Though more complex and expensive than co-precipitation or pyrolysis-based methods, the sol-gel approach offers advantages in chemical homogeneity, high purity, and the tailoring of functional surfaces for specific chemical or magnetic properties. Table 2.1 shows the magnetic precipitation, impregnation, hydrothermal synthesis, and sol-gel methods, along with their key advantages and disadvantages.

2.3   Factors Influencing Synthesis

The synthesis of magnetic biochar is significantly influenced by several key factors that directly affect the formation, dispersion, and performance of magnetic nanoparticles (typically Fe_3O_4 or $\gamma\text{-}Fe_2O_3$) on the biochar matrix. The most critical factors are described in the following sections.

2.3.1   Type of iron salt used

Commonly employed salts include ferric chloride ($FeCl_3$), ferrous sulphate ($FeSO_4$), and ferric nitrate ($Fe(NO_3)_3$). Each salt has different solubility, decomposition temperature, and reactivity, which influence the size, oxidation state, and magnetic strength of the formed iron oxides (Kucheryavy *et al.*, 2013). For example, $FeCl_3$ often leads to well-dispersed magnetite particles due to its high

Table 2.1. Magnetic precipitation, impregnation, hydrothermal synthesis, and sol-gel methods, with their key advantages and disadvantages.

Synthesis method	Description	Advantages	Disadvantages
Magnetic precipitation	Mixing iron salts (Fe^{2+}/Fe^{3+}) with biochar and adding base (e.g. NaOH/NH_4OH) to form Fe_3O_4 particles	Simple and cost-effective Good magnetic properties Environmentally friendly	Less control over particle size May require post-treatment Surface agglomeration
Impregnation	Biochar is soaked in metal salt solution (e.g. $FeCl_3$), then dried and thermally treated (Cho *et al.*, 2017)	Uniform dispersion possible Simple setup Low cost	Weak bonding of metal to support Metal leaching possible Non-uniform particle size
Hydrothermal synthesis	Reaction of precursors in a sealed autoclave at high temp and pressure	Produces crystalline, well-dispersed nanoparticles High-purity materials	Requires expensive equipment Time-consuming Not energy efficient
Sol-gel method	Formation of a colloidal solution (sol) that transitions into a gel, followed by drying/calcination (Schwartz, 1989)	High purity and homogeneity Control over structure and composition	Complex and time-intensive May involve expensive precursors Gel cracking possible

solubility, while $Fe(NO_3)_3$ may enhance porosity due to nitrate decomposition gases during pyrolysis.

2.3.2   Temperature and duration of the synthesis process

Pyrolysis or co-precipitation temperatures typically range from 400°C to 700°C; lower temperatures may produce amorphous or less magnetic oxides, while excessively high temperatures can cause particle agglomeration, reducing surface area and adsorption capacity. Longer reaction times can improve particle crystallinity but may also promote sintering if not controlled.

2.3.3 pH of the reaction medium

The pH is especially important in co-precipitation methods, where pH values between 9 and 11 are optimal for precipitating magnetite. Lower pH may result in incomplete precipitation or the formation of non-magnetic iron hydroxides, while higher pH may lead to unwanted phases (Bhakar *et al.*, 2025).

2.3.4 The ratio of iron to biochar or biomass

This ratio also plays a key role. A higher iron content typically enhances magnetization and adsorption capability, but excessive iron can block pores, reduce surface area, or result in aggregation of magnetic particles, lowering overall effectiveness. Finding the right balance is essential for optimizing surface functionality and recyclability.

2.3.5 Surface modification treatments

Surface modification treatments such as acid washing (with HCl or HNO_3) or alkaline activation (with NaOH or KOH) can improve the porosity and functional group density of the biochar, allowing better interaction with iron ions during synthesis and enhancing the dispersion and anchoring of magnetic particles (Amstad *et al.*, 2011). Acid washing can also remove unwanted ash or minerals, while base activation introduces more reactive –OH groups for improved particle bonding. Together, these factors must be carefully optimized based on the desired application to ensure the magnetic biochar has the appropriate surface area, porosity, magnetization strength, and chemical stability (Hao *et al.*, 2018).

2.4 Characterization Techniques for Magnetic Biochar

The characterization of magnetic biochar is important to validate its successful synthesis and to evaluate the structural, chemical, and magnetic properties that influence its performance in environmental and industrial applications. The process of characterization begins with confirming the presence of magnetic particles, typically Fe_3O_4, using X-ray diffraction (XRD) to identify their crystalline phases. Fourier-transform infrared spectroscopy (FTIR) is employed to analyse functional groups on the biochar surface that improve pollutant adsorption and interact with iron oxides (Liu *et al.*, 2023b). Scanning electron microscopy (SEM) and transmission electron microscopy (TEM) provide insights into the surface morphology, distribution of magnetic nanoparticles, and structural integration, while energy-dispersive X-ray spectroscopy (EDX) confirms elemental composition, especially iron content. To evaluate porosity and surface area, Brunauer–Emmett–Teller (BET) analysis is used, which is crucial for assessing the adsorptive potential of the material. Additionally, vibrating sample magnetometer (VSM) quantifies the magnetic behaviour, such as saturation magnetization, which determines how efficiently the biochar can be regenerated using external magnets. These characterization techniques collectively determine the desired surface chemistry, porosity, morphology, and magnetization for targeted applications like dye removal, heavy metal adsorption, or catalytic processes. The characterization of magnetic biochar using advanced analytical techniques is important for verifying its successful synthesis, understanding its physicochemical properties, and evaluating its effectiveness in applications such as wastewater treatment, catalysis, and environmental remediation. Each characterization technique provides specific and relevant information about the structure, composition, morphology, magnetic behaviour, thermal stability, and surface chemistry of magnetic biochar (Song *et al.*, 2021).

2.4.1   Fourier transform infrared spectroscopy (FTIR)

FTIR work on the principle that molecular bonds absorb infrared radiation at specific frequencies in respect to their vibrational modes. When IR radiation passes through a sample, certain wavelengths are absorbed by chemical bonds, producing a spectrum that reflects the molecular structure of the sample. FTIR is used to identify surface functional groups such as hydroxyl (–OH), carboxyl (–COOH), carbonyl (C=O), and metal–oxygen bonds like Fe–O. These groups are important in determining adsorption mechanisms (Lunardi *et al.*, 2021). For magnetic biochar, FTIR confirms the presence of Fe–O bonds, indicating successful magnetization, and evaluates changes in surface chemistry before and after modification or pollutant adsorption (Gao *et al.*, 2023).

2.4.2   X-ray diffraction (XRD)

XRD is based on the principle of constructive interference of monochromatic X-rays scattered by the crystal lattice of a material. The resulting diffraction pattern generates information about the crystallographic structure, phase identification, and particle size. XRD helps to identify the crystalline phases of magnetic components like magnetite (Fe_3O_4). The peaks at specific 2θ angles confirm the formation of these phases. XRD also confirm the crystal size using the Scherer equation and assesses phase purity, which is important for ensuring uniform magnetic behaviour (Zak *et al.*, 2012). The presence of sharp, well-defined peaks indicates successful iron oxide formation.

2.4.3   Scanning electron microscopy (SEM) and energy dispersive X-ray spectroscopy (EDS)

SEM uses a focused beam of electrons to scan the surface of a sample. The interaction of electrons with the surface atoms generates secondary and backscattered electrons, producing high-resolution images of surface features. EDS is typically coupled with SEM and detects X-rays emitted from the sample during electron bombardment. These X-rays are characteristic of specific elements, allowing for SEM to reveal the surface morphology – such as porosity, roughness, and distribution of magnetic nanoparticles – while EDS confirms the presence and distribution of iron within the biochar matrix (Neeli and Ramsurn, 2018). This combined technique is crucial for validating the homogeneous dispersion of magnetic particles and assessing potential agglomeration or structural defects.

2.4.4   Transmission electron microscopy (TEM)

TEM transmits a beam of electrons through an ultra-thin sample. The transmitted electrons are captured to form images with atomic or near-atomic resolution, revealing internal microstructures and fine features. TEM provides detailed information about the size, shape, and distribution of iron nanoparticles at the nanometre scale. It helps distinguish between magnetite and other iron oxides based on morphology and lattice fringes (Itoh and Sugimoto, 2003). TEM is particularly useful for observing core-shell structures, verifying whether iron particles are embedded in or coated on the biochar matrix.

2.4.5   Vibrating sample magnetometer (VSM)

In VSM, a sample is vibrated in a uniform magnetic field, inducing a voltage in nearby pickup coils. The magnitude of the voltage corresponds to the magnetic moment of the sample, which is plotted against the applied magnetic field. VSM evaluates key magnetic properties such as saturation magnetization (Ms), coercivity (Hc), and remanence (Mr). High saturation magnetization indicates effective magnetic recovery using an external magnet (Jia *et al.*, 2023). Low coercivity and remanence values confirm superparamagnetism, which is desirable for applications requiring easy separation and dispersion, such as in adsorption and catalysis.

2.4.6   Brunauer–Emmett–Teller (BET) surface area analysis

BET analysis is based on gas adsorption (usually nitrogen) at cryogenic temperatures. The amount of gas adsorbed is used to calculate the specific surface area and pore volume of the material. BET provides crucial insights into the adsorptive capacity of magnetic biochar by measuring its surface area, pore size distribution, and total porosity (Zhang *et al.*, 2020). A high surface area with mesoporous structure enhances adsorption of pollutants like dyes, heavy metals, and organic compounds. Surface area often decreases slightly after magnetization, which is also assessed via BET.

2.4.7 Thermogravimetric analysis (TGA)

TGA measures the change in the mass of a sample as it is heated under a controlled atmosphere (air or inert gas). It provides data on decomposition, thermal stability, and composition. TGA is used to assess the thermal stability of magnetic biochar and estimate the iron oxide content (Hu *et al.*, 2017). Residual mass at high temperatures (typically above 600–700°C) is attributed to stable inorganic residues, including Fe_3O_4. TGA also helps monitor moisture loss, volatile release, and degradation of organic matter, important for evaluating regeneration behaviour during repeated adsorption cycles.

2.4.8 Zeta potential analysis

Zeta potential measures the electrical potential at the slipping plane of particles suspended in a fluid. It reflects the degree of repulsion or attraction between particles, indicating colloidal stability. Table 2.2 shows the techniques used for magnetic biochar characterization.

2.5 Application of Techniques to Magnetic Biochar

Zeta potential analysis provides information about the surface charge of magnetic biochar across a range of pH levels. This is critical for understanding the electrostatic interaction between biochar and charged pollutants such as cationic dyes (e.g. methylene blue) or anionic compounds (e.g. phosphate, Congo red). Zeta potential also predicts dispersion stability and aids in optimizing pH for adsorption experiments (Lunardi *et al.*, 2021). Magnetic biochar characterized through these methods are being applied in wastewater treatment, catalysis, CO_2 capture, and even energy storage.

2.6 Conclusion

This chapter describes the synthesis and characterization of magnetic biochar as an efficient, magnetically separable, and reusable adsorbent for wastewater treatment. It covers key synthesis methods – such as co-precipitation, pyrolysis-based techniques, hydrothermal, and sol-gel approaches – and stresses the importance of controlling reaction parameters to tailor performance. Characterization techniques such as FTIR, XRD, SEM-EDS, TEM, VSM, BET, TGA, and zeta potential are essential to validate magnetization and understand the structure, surface chemistry, and adsorption potential of biochar. The integration of well-designed synthesis and thorough characterization is important for optimizing magnetic biochar for scalable environmental applications.

Table 2.2. Techniques used for magnetic biochar characterization.

Technique	Working principle (brief)	Application in magnetic biochar
FTIR (Fourier transform infrared spectroscopy)	Measures absorption of infrared light to detect molecular vibrations and identify functional groups	Identifies surface groups like –OH, –COOH, Fe–O Confirms surface modification and bonding with iron
XRD (X-ray diffraction)	Analyses diffraction patterns from crystal planes when exposed to X-rays (Aguilar-Marín *et al.*, 2020)	Detects crystalline phases (e.g. Fe_3O_4, γ-Fe_2O_3) Determines crystallite size and phase purity
SEM (scanning electron microscopy)	Uses focused electron beams to produce detailed surface images	Observes surface morphology, pore structure, and particle dispersion
EDS (Energy dispersive X-ray spectroscopy)	Detects characteristic X-rays emitted by elements under electron beam excitation (Goldstein *et al.*, 2017)	Confirms elemental composition (presence of Fe, C, O, etc.)
TEM (transmission electron microscopy)	Passes electrons through thin samples to image fine structures	Reveals internal structure and iron nanoparticle distribution Nanoparticle size and shape analysis
VSM (Vibrating sample magnetometer)	Measures magnetization as a sample vibrates in a magnetic field	Quantifies magnetic properties like saturation magnetization and coercivity Identifies super paramagnetism
BET (Brunauer–Emmett–Teller) surface area analysis	Uses gas adsorption to measure specific surface area and porosity	Determines surface area and pore size critical for adsorption efficiency
TGA (thermogravimetric analysis)	Measures mass change of a sample with temperature	Assesses thermal stability Estimates iron oxide content via residual weight
Zeta potential analysis	Measures the electrostatic potential at the slipping plane of particles in a liquid (Anderson, 1994)	Indicates surface charge Predicts colloidal stability and adsorption affinity

References

Aguilar-Marín, P., Angelats-Silva, L., Noriega-Diaz, E., Chavez-Bacilio, M. and Verde-Vera, R. (2020) Understanding the phenomenon of X-ray diffraction by crystals and related concepts. *European Journal of Physics* 41(4), 045501. DOI: 10.1088/1361-6404/ab8e53.

Almeida, T.P. (2010) Hydrothermal synthesis and characterisation of α-Fe2O3 nanorods. Doctoral dissertation, University of Nottingham, Nottingham, UK.

Amstad, E., Textor, M. and Reimhult, E. (2011) Stabilization and functionalization of iron oxide nanoparticles for biomedical applications. *Nanoscale* 3(7), 2819–2843. DOI: 10.1039/c1nr10173k.

Anderson, R.A. (1994) Electrostatic forces in an ideal spherical-particle electrorheological fluid. *Langmuir* 10(9), 2917–2928. DOI: 10.1021/la00021a013.

Awang, N.A., Wan Salleh, W.N., Aziz, F., Yusof, N. and Ismail, A.F. (2023) A review on preparation, surface enhancement and adsorption mechanism of biochar-supported nano zero-valent iron adsorbent for hazardous heavy metals. *Journal of Chemical Technology & Biotechnology* 98(1), 22–44. DOI: 10.1002/jctb.7182.

Bhakar, S.K., Sharma, B., Satapathy, S., Deshmukh, P., Srihari, V. *et al.* (2025) Optimizing synthesis parameters to achieve phase-pure superparamagnetic Fe_3O_4 nanoparticles for magnetic hyperthermia. *Bio Nano Science* 15(3), 1–20. DOI: 10.1007/s12668-025-02130-y.

Bramsiepe, C., Sievers, S., Seifert, T., Stefanidis, G.D., Vlachos, D.G. *et al.* (2012) Low-cost small scale processing technologies for production applications in various environments—mass produced factories. *Chemical Engineering and Processing* 51, 32–52. DOI: 10.1016/j.cep.2011.08.005.

Cho, D.W., Yoon, K., Kwon, E.E., Biswas, J.K. and Song, H. (2017) Fabrication of magnetic biochar as a treatment medium for as (V) via pyrolysis of $FeCl_3$-pretreated spent coffee ground. *Environmental Pollution (Barking, Essex)* 229, 942–949. DOI: 10.1016/j.envpol.2017.07.079.

Cho, K.Y., Seo, H.Y., Yeom, Y.S., Kumar, P., Lee, A.S. *et al.* (2016) Stable 2D-structured supports incorporating ionic block copolymer-wrapped carbon nanotubes with graphene oxide toward compact decoration of metal nanoparticles and high-performance nano-catalysis. *Carbon* 105, 340–352. DOI: 10.1016/j.carbon.2016.04.049.

de Souza Mesquita, L.M., Contieri, L.S., Silva, F.A., Bagini, R.H., Bragagnolo, F.S. *et al.* (2024) Path2Green: Introducing 12 green extraction principles and a novel metric for assessing sustainability in biomass valorization. *Green Chemistry* 26(19), 10087–10106. DOI: 10.1039/d4gc02512a.

Gao, G., Zhou, P., Chen, C. and Zhu, L. (2023) Adsorption of MB and Pb(II) before and after magnetic modification: Performance and mechanism. *Journal of Molecular Structure* 1293, 136306. DOI: 10.1016/j.molstruc.2023.136306.

Goldstein, J.I., Newbury, D.E., Michael, J.R., Ritchie, N.W., Scott, J.H.J. *et al.* (2017) X-rays. In: *Scanning Electron Microscopy and X-Ray Microanalysis*. Springer, New York, NY, pp. 39–63. DOI: 10.1007/978-1-4939-6676-9_4.

Guna, V., Ilangovan, M., Anantha Prasad, M.G. and Reddy, N. (2017) Water hyacinth: A unique source for sustainable materials and products. *ACS Sustainable Chemistry & Engineering* 5(6), 4478–4490. DOI: 10.1021/acssuschemeng.7b00051.

Hamamah, Z.A. and Grützner, T. (2022) Liquid-Liquid centrifugal extractors: Types and recent applications – a review. *Chemical and Biological Engineering Reviews* 9(3), 286–318. DOI: 10.1002/cben.202100035.

Hao, Z., Wang, C., Yan, Z., Jiang, H. and Xu, H. (2018) Magnetic particles modification of coconut shell-derived activated carbon and biochar for effective removal of phenol from water. *Chemosphere* 211, 962–969. DOI: 10.1016/j. chemosphere.2018.08.038.

Hu, X., Xu, J., Wu, M., Xing, J., Bi, W. *et al.* (2017) Effects of biomass pre-pyrolysis and pyrolysis temperature on magnetic biochar properties. *Journal of Analytical and Applied Pyrolysis* 127, 196–202. DOI: 10.1016/j.jaap.2017.08.006.

Huang, D., Wu, J., Wang, L., Liu, X., Meng, J. *et al.* (2019) Novel insight into adsorption and co-adsorption of heavy metal ions and an organic pollutant by magnetic graphene nanomaterials in water. *Chemical Engineering Journal* 358, 1399–1409. DOI: 10.1016/j.cej.2018.10.138.

Hunter, R.D. (2023) Synthesis of nanostructured graphitic carbons by iron-catalyzed graphitization of biomass. Doctoral dissertation, University of Birmingham, Birmingham, UK.

Hussain, S., Aneggi, E. and Goi, D. (2021) Catalytic activity of metals in heterogeneous Fenton-like oxidation of wastewater contaminants: A review. *Environmental Chemistry Letters* 19(3), 2405–2424. DOI: 10.1007/s10311-021-01185-z.

Itoh, H. and Sugimoto, T. (2003) Systematic control of size, shape, structure, and magnetic properties of uniform magnetite and maghemite particles. *Journal of Colloid and Interface Science* 265(2), 283–295. DOI: 10.1016/ s0021-9797(03)00511-3.

Jia, Z., Kou, M., Li, Y., Cao, S., Ding, G. *et al.* (2023) Effect of grain boundary reconstruction and regenerated main phase shell on magnetic properties in high-abundance (NdLaCeY)-Fe-B magnets. *Journal of Materials Research and Technology* 24, 4500–4509. DOI: 10.1016/j.jmrt.2023.04.110.

Kucheryavy, P., He, J., John, V.T., Maharjan, P., Spinu, L. *et al.* (2013) Superparamagnetic iron oxide nanoparticles with variable size and an iron oxidation state as prospective imaging agents. *Langmuir* 29(2), 710–716. DOI: 10.1021/la3037007.

Lan, Y., Gai, S., Cheng, K., Li, J. and Yang, F. (2022) Lanthanum carbonate hydroxide/ magnetite nanoparticles functionalized porous biochar for phosphate adsorption and recovery: Advanced capacity and mechanisms study. *Environmental Research* 214(Pt 1), 113783. DOI: 10.1016/j.envres.2022.113783.

Liu, Y., Li, C., Zhang, H., Fan, X., Liu, Y. *et al.* (2015) One-pot hydrothermal synthesis of highly monodisperse water-dispersible hollow magnetic microspheres and construction of photonic crystals. *Chemical Engineering Journal* 259, 779–786. DOI: 10.1016/j.cej.2014.08.051.

Liu, M., Ye, Y., Ye, J., Gao, T., Wang, D. *et al.* (2023a) Recent advances of magnetite (Fe$_3$O$_4$)-based magnetic materials in catalytic applications. *Magnetochemistry* 9(4), 110. DOI: 10.3390/magnetochemistry9040110.

Liu, Y., Wang, L., Liu, C., Ma, J., Ouyang, X. *et al.* (2023b) Enhanced cadmium removal by biochar and iron oxides composite: Material interactions and pore structure. *Journal of Environmental Management* 330, 117136. DOI: 10.1016/j. jenvman.2022.117136.

Lunardi, C.N., Gomes, A.J., Rocha, F.S., Tommaso, J. and Patience, G.S. (2021) Experimental methods in chemical engineering: Zeta potential. *The Canadian Journal of Chemical Engineering* 99(3), 627–639. DOI: 10.1002/cjce.23914.

Neeli, S.T. and Ramsurn, H. (2018) Synthesis and formation mechanism of iron nanoparticles in graphitized carbon matrices using biochar from biomass model compounds as a support. *Carbon* 134, 480–490. DOI: 10.1016/j. carbon.2018.03.079.

Rath, P., Jindal, M. and Jindal, T. (2021) A review on economically-feasible and environmental-friendly technologies promising a sustainable environment. *Cleaner Engineering and Technology* 5, 100318. DOI: 10.1016/j.clet.2021.100318.

Schwartz, R.W. (1989) *Chemical Processing of Lead Metatitanate by Co-Precipitation and Sol-Gel Methods: The Role of Powder and Gel Characteristics on Crystallization Behavior.* University of Illinois, Urbana-Champaign.

Song, J., Huang, Z. and Gamal El-Din, M. (2021) Adsorption of metals in oil sands process water by a biochar/iron oxide composite: Influence of the composite structure and surface functional groups. *Chemical Engineering Journal* 421, 129937. DOI: 10.1016/j.cej.2021.129937.

Torres-Gómez, N., Nava, O., Argueta-Figueroa, L., García-Contreras, R., Baeza-Barrera, A. *et al.* (2019) Shape tuning of magnetite nanoparticles obtained by hydrothermal synthesis: Effect of temperature. *Journal of Nanomaterials* 2019(1), 1–15. DOI: 10.1155/2019/7921273.

Tripathi, M., Mubarak, N., Sahu, J. and Ganesan, P. (2016) Overview on synthesis of magnetic bio char from discarded agricultural biomass. *Handbook of Composites from Renewable Materials, Structure and Chemistry* 1, 435. DOI: 10.1002/9781119441632.

Weidner, E., Karbassiyazdi, E., Altaee, A., Jesionowski, T. and Ciesielczyk, F. (2022) Hybrid metal oxide/biochar materials for wastewater treatment technology: A review. *ACS Omega* 7(31), 27062–27078. DOI: 10.1021/acsomega.2c02909.

Zak, A.K., Majid, W.H., Abrishami, M.E., Yousefi, R. and Parvizi, R. (2012) Synthesis, magnetic properties and X-ray analysis of ZnO. 97X0. 03O nanoparticles (X= Mn, Ni, and Co) using Scherrer and size–strain plot methods. *Solid State Sciences* 14(4), 488–494. DOI: 10.1016/j.solidstatesciences.2012.01.019.

Zhang, N., Reguyal, F., Praneeth, S. and Sarmah, A.K. (2023) A green approach of biochar-supported magnetic nanocomposites from white tea waste: Production, characterization and plausible synthesis mechanisms. *Science of The Total Environment* 886, 163923. DOI: 10.1016/j.scitotenv.2023.163923.

Zhang, P., O'Connor, D., Wang, Y., Jiang, L., Xia, T. *et al.* (2020) A green biochar/iron oxide composite for methylene blue removal. *Journal of Hazardous Materials* 384, 121286. DOI: 10.1016/j.jhazmat.2019.121286.

Ziemniak, S.E., Jones, M.E. and Combs, K.E.S. (1995) Magnetite solubility and phase stability in alkaline media at elevated temperatures. *Journal of Solution Chemistry* 24(9), 837–877. DOI: 10.1007/BF00973442.

Properties and Applications of Magnetic Biochar in Wastewater Treatment

3

3.1 Introduction

Water pollution is considered as one of the most serious global environmental challenges of the 21st century. Rapid industrialization, urbanization, and population growth have meaningfully increased the discharge of pollutants into water bodies, including organic dyes, heavy metals, pharmaceuticals, pesticides, and additional nutrients like nitrates and phosphates (Velusamy *et al.*, 2021). These pollutants not only degrade aquatic ecosystems but also pose risks to human health and food safety. Industrial sectors such as textiles, pharmaceuticals, electroplating, agriculture, and domestic waste contribute to the generation of untreated or partially treated wastewater, thereby compounding the problem. Conventional wastewater treatment methods – such as coagulation–flocculation, sedimentation, biological treatment, and membrane filtration – have been widely employed for pollutant removal (Zinicovscaia, 2016). However, these methods often face major problems such as high operational costs, the generation of secondary sludge, limited effectiveness against initial contaminants, and energy-intensive processes, which makes them less sustainable in the long term. Many of these approaches are not efficient for treating low-concentration pollutants or for achieving selective pollutant removal (Lyu *et al.*, 2014).

These challenges have driven the need for alternative, cost-effective, and environmentally friendly technologies that can deliver high performance with lower energy input and waste generation. In recent years, biochar has drawn significant attention as an effective adsorbent in wastewater treatment. Biochar is a carbon-rich, porous material derived from biomass pyrolysis under limited oxygen conditions (Devi *et al.*, 2023). Its porous structure, large surface area, and surface functional groups make it suitable for capturing a variety of pollutants through adsorption. Despite its advantages, conventional

Corresponding author: danbahadur.chem@gmail.com

© CAB International 2026. *Magnetic Biochar for Wastewater Remediation in the Textile Industry* (D. Bahadur Pal and A. Kapoor)

biochar often suffers from drawbacks such as poor selectivity, limited reusability, and difficulty in separation from treated water (Shakoor *et al.*, 2020). To solve this problem, researchers have developed magnetic biochar by integrating magnetic particles – primarily iron-based materials like Fe_3O_4 or γ-Fe_2O_3 – into the biochar matrix. This modification not only enhances the adsorptive performance through improved surface functionality and reactivity but also provides magnetic properties, permitting easy separation from aqueous media using external magnetic fields (Manyangadze *et al.*, 2020). Magnetic biochar has improved reusability, reduces the need for chemical coagulants or filters, and fits well with sustainable and circular economy principles.

The purpose of this chapter is to deliver a comprehensive overview of the properties and applications of magnetic biochar in wastewater treatment. The chapter examines the physical, chemical, magnetic, and mechanical characteristics that influence its adsorption performance, such as Brunauer–Emmett–Teller (BET) surface area, surface functional groups, magnetization behaviour, and structural stability (Ramos Guivar *et al.*, 2017). This chapter also describes various real-world applications of magnetic biochar in removing both organic and inorganic contaminants, including dyes, heavy metals, pharmaceuticals, and nutrients. There is a critical evaluation of the advantages of magnetic biochar over conventional adsorbents, the techniques for its regeneration and reuse, and its performance in pilot-scale and real-world scenarios (Srai *et al.*, 2016). It concludes by discussing future prospects, including innovations in low-cost synthesis, hybrid systems integration, and functionalization strategies to improve performance. This comprehensive discussion aims to provision ongoing research and industrial adoption of magnetic biochar as a sustainable and efficient material for advanced wastewater treatment (Meng *et al.*, 2025).

3.2   Properties of Magnetic Biochar

There are number of physical properties – high BET surface area, multimodal porosity, optimal particle size morphology, and efficient magnetism – that enhance the adsorption performance and process feasibility of magnetic biochar (Dong *et al.*, 2022). These properties not only increase contaminant capture but also simplify downstream processing through magnetic retrieval, supporting the development of sustainable, cost-effective, and high-performance wastewater treatment technologies.

3.2.1   Physical properties

The physical properties of magnetic biochar play a vital role in determining its efficiency and applicability in wastewater treatment. Among these, surface area and porosity are crucial characteristics.

3.2.1.1   Surface area

The BET method provides insight into the number of available active sites for adsorption. A higher BET surface area directly correlates with enhanced pollutant uptake, making it a key indicator of performance adsorbent.

3.2.1.2   Porosity

The porosity of biomass refers to the pore size distribution within the biochar structure, which is usually categorized into microspores (<2 nm), mesopores (2–50 nm), and macrospores (>50 nm) (Mays, 2007). Each pore type contributes differently to adsorption efficiency: microspores provide high surface area and are ideal for trapping small molecules like heavy metals and dyes; mesopores support the diffusion of pollutants into internal layers, facilitating faster adsorption kinetics; and macrospores serve as channels for fluid movement, improving mass transfer (Roper and Seminara, 2019). Together, a hierarchical pore structure enhances both the capacity and rate of contaminant removal by offering multiple mechanisms such as multilayer adsorption, pore-filling, and electrostatic interaction. In addition to porosity, particle size and morphology significantly influence the biochar's interaction with pollutants. Smaller particle sizes offer a larger surface-to-volume ratio and reduce diffusion distances for pollutants, thereby increasing the rate of adsorption. However, excessively small particles can tend to aggregate, which may reduce effective surface area and hinder separation processes.

3.2.1.3   Morphology

Imaging techniques such as scanning electron microscopy (SEM) reveal surface texture and structural features. Magnetic biochar with a rough, porous, and irregular surface morphology generally has more adsorption sites and improved contact with contaminants in water. Such features enable enhanced adsorption efficiency by increasing the likelihood of pollutant–biochar interactions (Hu *et al.*, 2020). Thus, controlling particle size and developing favourable morphological features are essential for achieving optimal adsorption performance.

3.2.1.4   Magnetism

Magnetism is imparted through the inclusion of magnetic materials like Fe_3O_4 (magnetite) or $\gamma\text{-}Fe_2O_3$ (magnetite). This introduces the ability to separate the material from aqueous media using an external magnetic field, eliminating the need for filtration or centrifugation. Magnetism in biochar is often characterized by two behaviours: superparamagnetism and ferromagnetism (Khan *et al.*, 2015). Superparamagnetic materials exhibit magnetic properties only when an external field is applied and lose magnetization once the field is removed, thus preventing unwanted aggregation in solution. In contrast, ferromagnetic materials retain their magnetization, which may complicate dispersion but enhance separation efficiency.

3.2.2   Chemical properties

The chemical properties of magnetic biochar play a critical role in its effectiveness as an adsorbent for wastewater treatment.

3.2.2.1   Functional group

Surface functional groups such as carboxyl (–COOH), hydroxyl (–OH), and carbonyl (C=O) are particularly important. These groups are typically introduced during the pyrolysis process or through post-treatment modifications and are responsible for the majority of interactions with pollutants (Tomei *et al.*, 2019). These functional groups facilitate adsorption through mechanisms like hydrogen bonding, electrostatic attraction, surface complexation, and ion exchange. For example, in the removal of cationic dyes such as methylene blue, negatively charged functional groups (especially carboxyl and hydroxyl) on the magnetic biochar surface attract and bind positively charged dye molecules via electrostatic interaction (Hu *et al.*, 2013). In contrast, in the case of heavy metal adsorption, these functional groups act as ligands that coordinate with metal ions to form stable surface complexes, improving both the capacity and selectivity of the adsorbent.

3.2.2.2   Energy dispersive X-ray spectroscopy (EDS) and CHNS analysis

Another essential chemical property is the elemental composition of magnetic biochar, typically analysed using energy dispersive X-ray spectroscopy (EDS) or CHNS analysis (carbon, hydrogen, nitrogen, and sulphur) These techniques help quantify the elemental makeup of the material, especially the contents of carbon, oxygen, and iron. High carbon content reflects the presence of a stable carbon matrix with extensive aromatic structures, which enhances mechanical and thermal stability. Oxygen-rich functional groups contribute to hydrophilicity and reactivity, thereby improving adsorption affinity toward polar contaminants (Xia and Ball, 1999). The presence of iron, introduced through impregnation or co-precipitation of iron salts, imparts magnetism and contributes to redox reactions or Fenton-like catalytic activity in advanced oxidation processes (AOPs). Elemental composition analysis helps assess the chemical integrity of the material before and after pollutant adsorption, and during regeneration cycles, ensuring long-term usability (Smidt and Lechner, 2005).

3.2.2.3   pH PZC (point of zero charge)

The pH PZC is another significant parameter that influences the surface charge of magnetic biochar under varying pH conditions. It is the pH at which the biochar surface has a neutral charge. When the solution pH is above the pH PZC, the surface becomes negatively charged, favouring adsorption of cationic pollutants such as metal ions or basic dyes. Conversely, when the pH is below the pH PZC, the surface becomes positively charged, which increase the adsorption of anionic pollutants like nitrate, phosphate, or acidic dyes. The adjusting the solution pH in relation to pH PZC allows for targeted removal of specific contaminants and helps optimize adsorption conditions in practical applications.

3.2.2.4   Thermogravimetric analysis (TGA)

The thermal stability adsorbent typically estimated through thermogravimetric analysis (TGA) is a key factor that determines the durability and reusability of magnetic biochar. TGA measures the weight loss of the material as a function of temperature, providing information about its decomposition profile and stability (Smidt and Lechner, 2005). A biochar with high thermal stability can withstand multiple adsorption–desorption cycles without significant structural degradation, which is crucial for economic and operational sustainability. Moreover, thermal stability is vital when regeneration is performed using thermal methods, as the adsorbent must endure elevated temperatures without loss of functionality or magnetism (Kawamura, 2000).

3.2.3   Mechanical and structural properties

The durability, structural integrity, and low agglomeration tendency of magnetic biochar are vital for its repeated use and scalability in wastewater treatment. These properties ensure long-term efficiency, easy handling, and practical arrangement in diverse water treatment settings (Kawamura, 2000). When optimized, they make magnetic biochar a reliable and cost-effective material for sustainable environmental remediation. The mechanical and structural properties of magnetic biochar are essential for measuring its practical viability in long-term and large-scale wastewater treatment processes. These properties affect how the material performs under repeated use, its stability during application, and its compatibility with aqueous systems.

3.2.3.1   Durability and reusability

Durability refers to the material's ability to preserve its physical and chemical properties over time and repeated operational cycles. In the context of magnetic biochar, reusability is a key factor that defines its cost-effectiveness and environmental sustainability (Psarommatis and May, 2025). An ideal adsorbent should resist several adsorption–desorption cycles without significant loss of adsorption capacity or magnetic properties. This resistance to degradation is measured through repeated use in batch or continuous systems. Magnetic biochar typically demonstrates high resistance to mechanical breakdown and retains its structural integrity due to the carbonaceous backbone formed during pyrolysis (Devi *et al.*, 2020). Additionally, the iron oxide components (e.g. Fe_3O_4) used for magnetization are generally stable under mild regeneration conditions (acid/base washing, magnetic separation), allowing the material to be reused efficiently. This regenerative ability reduces the need for frequent replacement and lowers operational costs.

3.2.3.2 *Structural integrity after magnetization*

The structural integrity of magnetic biochar refers to its ability to maintain a stable porous structure and functional performance after being modified with magnetic materials. Compared to pristine biochar (unmodified), magnetic biochar often exhibits some changes in texture, porosity, and mechanical strength due to the deposition of iron oxides on its surface or within its pores (Tartaj and Amarilla, 2011). However, properly controlled synthesis techniques – such as co-precipitation or impregnation pyrolysis – can preserve the essential pore structure and minimize collapse or blockage. The addition of magnetic components may sometimes reduce overall surface area or change the hydrophobicity/hydrophilicity of the biochar, but these effects can be balanced by enhanced separation ability and multifunctionality (Long *et al.*, 2017). Studies have shown that, with optimized synthesis, magnetic biochar retains comparable or even improved structural robustness and adsorption performance relative to pristine biochar, especially in high-load or industrial wastewater scenarios (Gonçalves *et al.*, 2025).

3.2.3.3 *Agglomeration tendency*

Agglomeration is the tendency of particles to clump together, particularly in aqueous environments. In magnetic biochar, this behaviour can be influenced by particle size, surface charge, and magnetic interactions. If not properly addressed, agglomeration can hinder dispersibility, reduce the available surface area for adsorption, and limit contact between the adsorbent and pollutants (Akhtar *et al.*, 2024). This affects the adsorption kinetics and efficiency of the material. Superparamagnetic behaviour achieved by maintaining nanoparticle sizes below a critical threshold helps reduce magnetic dipole–dipole attraction among particles, thus minimizing agglomeration. Additionally, surface modification with stabilizing agents (e.g. humic acid, polymers) or maintaining proper pH and ionic strength in solution can further reduce particle clumping. Ensuring good dispersion in water enhances the contact area between biochar and pollutants, making the adsorption process more efficient and predictable (Haider Jaffari *et al.*, 2023).

3.3   Applications of Magnetic Biochar in Wastewater Treatment

3.3.1   Removal of organic pollutants using magnetic biochar

The discharge of organic pollutants such as synthetic dyes, pharmaceuticals, pesticides, and herbicides into natural water bodies from industrial, agricultural, and domestic sources has become a serious environmental concern. These contaminants are often persistent, toxic, and bio-accumulative, posing risks to aquatic ecosystems and human health. Conventional wastewater treatment methods are generally inadequate for effectively removing such organic micropollutants due to their stable molecular structures and low bio-degradability. In this context, magnetic biochar has emerged as a highly effective, reusable, and eco-friendly adsorbent for removing organic contaminants from wastewater (Akhtar *et al.*, 2024). Its large surface area, tailored surface chemistry, porous structure, and magnetic properties make it particularly suitable for wastewater treatment applications. This discussion focuses on three major categories of organic pollutants: dyes, pharmaceutical compounds, and pesticides/herbicides, and the mechanisms through which magnetic biochar facilitates their removal.

3.3.2   Dye adsorption

Synthetic dyes are among the most prevalent pollutants in industrial wastewater, especially from the textile, leather, paper, and food industries. These dyes are often toxic, carcinogenic, and resistant to degradation (Ismail *et al.*, 2019). Common examples include methylene blue (MB), Congo red (CR), and Rhoda mine B (RhB), which serve as model dyes in adsorption studies. Magnetic biochar has demonstrated excellent capability in adsorbing such dyes due to its favourable surface characteristics and functional groups (Li *et al.*, 2022). The primary mechanisms involved in dye adsorption include electrostatic interactions. For example, MB is a cationic dye, while CR is an anionic dye. Magnetic biochar surfaces typically contain negatively charged groups (e.g. carboxyl, hydroxyl) when the solution pH is above the pH PZC, facilitating electrostatic attraction of positively charged dye molecules. Similarly, at low pH, protonated surfaces can attract anionic dyes. Another mechanism is π–π stacking interactions. Biochar, being rich in aromatic structures due to its carbonized matrix, allows for π–π interactions with dye molecules that also contain aromatic rings (e.g. RhB, CR) (Gęca *et al.*, 2023). These interactions are non-covalent and significantly enhance dye uptake, especially for compounds with delocalized π-electron systems. Pore filling is a third dye adsorption mechanism for magnetic biochar. The hierarchical porous structure of magnetic biochar comprising micro-, meso-, and macropores offers multiple pathways for dye molecules to enter and occupy adsorption sites (Lu *et al.*, 2020). Micropores provide the surface area, mesopores facilitate transport, and macropores serve as entry channels, altogether optimizing dye removal. Several studies have reported over 90% removal efficiency for MB and CR using magnetically modified biochar derived from biomass such as water hyacinth, rice husk, or coconut shell. The reusability of magnetic biochar enabled by simple magnetic separation and regeneration makes it an economically viable adsorbent for dye removal (Li *et al.*, 2022).

3.3.3 Pharmaceutical pollutants

Pharmaceutical residues including antibiotics, analgesics, hormones, and anti-inflammatory drugs are classified as emerging organic pollutants (EOPs). These compounds enter water systems through hospital effluents, household waste, and improper disposal. They are persistent, biologically active, and capable of disrupting endocrine systems and contributing to antibiotic resistance (Guarnotta *et al.*, 2022). Magnetic biochar has shown promise in the removal of pharmaceutical pollutants such as tetracycline and ciprofloxacin, due to its multifunctional surface chemistry. The adsorption mechanisms in this case are slightly more complex and include:

1. **Hydrogen bonding**: Functional groups on the biochar surface (e.g. –OH, –COOH) can form hydrogen bonds with polar functional groups on pharmaceuticals, enhancing retention.
2. **Electrostatic attraction/repulsion**: Depending on the pH of the solution and the charge of the pharmaceutical compound, magnetic biochar can attract or repel specific drugs. For instance, tetracycline carries a net positive charge under acidic conditions and can be effectively adsorbed onto negatively charged surfaces.
3. **Hydrophobic interaction**: Nonpolar regions of pharmaceutical molecules can adsorb onto hydrophobic domains on the biochar, particularly those formed during high-temperature pyrolysis.
4. **π–π electron donor–acceptor interactions**: Many pharmaceutical compounds and biochar surfaces contain conjugated systems that allow for π–π stacking, enhancing adsorption efficiency (Tang *et al.*, 2022).

Several experiments have reported removal efficiencies exceeding 80–95% for antibiotics like ciprofloxacin and tetracycline using iron-modified magnetic biochars. The adsorption is typically fast, and the adsorbent can be regenerated through mild acid washing or solvent desorption, maintaining performance over multiple cycles.

3.3.4 Pesticides and herbicides

Pesticides and herbicides such as atrazine, glyphosate, diazinon, and chlorpyrifos are heavily used in agriculture and frequently detected in surface and groundwater (Jacob *et al.*, 2024). They are persistent organic pollutants

(POPs) that can cause neurological, reproductive, and carcinogenic effects in humans and wildlife. Magnetic biochar provides a cost-effective and efficient solution for removing these pollutants. Its high porosity, active surface groups, and magnetism make it suitable for adsorption in field and industrial applications (Zeng *et al.*, 2022). The main removal mechanisms include:

1. **Hydrophobic partitioning**: Many pesticides are hydrophobic and can be adsorbed onto the hydrophobic surfaces of magnetic biochar, especially those prepared at high pyrolysis temperature.
2. **Electrostatic interaction and ligand exchange**: Pesticides with ionizable functional groups can interact electrostatically with surface groups or undergo ligand exchange with Fe–OH or Fe–O sites on the biochar.
3. **Pore trapping and van der Waals forces**: The micropores and mesopores of magnetic biochar trap pesticide molecules through weak van der Waals interactions, leading to physical entrapment.

3.3.5 Case studies

A study has shown that iron-oxide-impregnated biochar derived from agricultural waste (e.g. wheat straw, corn cob) can achieve up to 90% removal of atrazine from contaminated water (Siddiqui *et al.*, 2023). Moreover, real-field samples from irrigation runoff have confirmed the efficacy of magnetic biochar in binding and immobilizing a range of pesticide compounds, thus preventing them from entering aquatic ecosystems (Riaz *et al.*, 2021).

Magnetic biochar has shown significant potential in removing inorganic pollutants such as heavy metals and nutrients from wastewater (Rezania *et al.*, 2015). The magnetic property enables easy separation after treatment and enhances adsorption through mechanisms such as ion exchange, surface complexion, and precipitation. Magnetic biochar effectively removes toxic heavy metals such as Pb^{2+}, Cd^{2+}, Cr (VI), and As (III). For example, Pb^{2+} and Cd^{2+} are seized through exchange with surface ions and coordination with functional groups, while Cr (VI) may be minimized and precipitated as Cr $(OH)_3$ on the magnetic biochar surface. Studies have reported high removal efficiencies (>95%) for Cr (VI) and Pb^{2+} using Fe-doped biochar derivative from biomass like water hyacinth and rice husk (Long *et al.*, 2017) . In addition, magnetic biochar plays a crucial role in nutrient removal to combat eutrophication, particularly by removing phosphate and nitrate (Cheng *et al.*, 2022). Phosphate ions are captured through electrostatic attraction, ligand exchange with surface hydroxyls, or precipitation as metal phosphates with removal efficiencies often exceeding 90%. Though nitrate is more mobile and less interactive with carbon surfaces, magnetic biochar modified with redox-active metals can adsorb nitrates or facilitate their microbial denitrification. Overall, magnetic biochar delivers a multifunctional, reusable, and environmentally friendly platform for treating contaminated water capable of removing both heavy metals and nutrients simultaneously, with the added advantage of easy magnetic separation and regeneration potential (Nikić *et al.*, 2024). Its versatility and efficiency make it a promising material for sustainable water remediation applications.

3.3.6   Simultaneous removal of multiple pollutants

Magnetic biochar has advantages as an alternative to traditional adsorbents like activated carbon, conventional biochar, and metal-organic frameworks (MOFs) in its cost-effectiveness, ease of recovery, and environmental sustainability (see Table 3.1). It can be derived from low-cost biomass (such as agricultural or aquatic waste) and modified with iron salts through methods such as co-precipitation or impregnation, with a relatively low-to-moderate production cost, especially when compared to activated carbon and MOFs (Sharma and Kamalesu, 2023). Activated carbon, although known for its high surface area and excellent adsorption capacity, is expensive to produce due to its energy-intensive activation processes and the use of non-renewable precursors. Moreover, its regeneration is limited and disposal becomes a concern after saturation. In contrast, magnetic biochar not only provides comparable or even superior adsorption performance for heavy metals and nutrients but also adds a unique advantage of magnetic recovery. This feature significantly simplifies the separation process from treated water, making magnetic biochar

Table 3.1. A comparison of magnetic biochar with other adsorbents (Dong *et al.*, 2022).

Parameter	Magnetic biochar	Activated carbon	Conventional biochar	Metal–organic frameworks
Cost	Low to moderate (biomass + iron salts; scalable)	High (due to expensive activation processes)	Very low (simple pyrolysis of biomass)	Very high (costly precursors and synthesis techniques)
Ease of recovery	Excellent (magnetically separable)	Poor (requires filtration or centrifugation)	Poor (non-magnetic, difficult to recover)	Moderate (some are recoverable, but often delicate)
Environmental friendliness	High (renewable biomass + reusable + minimal waste)	Moderate (effective but energy-intensive production)	High (renewable, low impact, though less efficient)	Low to moderate (synthetic, may involve toxic solvents)

highly suitable for continuous or large-scale operations. In terms of ease of recovery, activated carbon and conventional biochar require filtration, sedimentation, or centrifugation – processes that are energy-consuming and often inefficient – whereas magnetic biochar can be quickly removed using an external magnetic field (Niu *et al.*, 2025). Compared to conventional biochar, which is inexpensive and environmentally benign, magnetic biochar retains these eco-friendly attributes whilst also offering enhanced adsorption due to the presence of iron oxides that introduce functional groups, increase surface area, and enable redox reactions. Conventional biochar is non-magnetic, making it harder to recover and reuse, which limits its practicality in dynamic water treatment systems (Gillingham, 2022). MOFs, on the other hand, are advanced crystalline materials known for their high porosity and tuneable properties, making them highly selective and efficient adsorbents. Nevertheless, their high synthesis cost, reliance on expensive and sometimes toxic solvents, and complex manufacturing methods make them less viable for large-scale or low-resource settings. Additionally, some MOFs are sensitive to moisture or pH changes, reducing their durability in real wastewater environments (Liu *et al.*, 2020). In terms of environmental friendliness, MBC ranks high due to its base in renewable biomass and reduced chemical footprint during synthesis, especially when green methods are used. Activated carbon is moderately sustainable but can contribute to carbon emissions if derived from fossil-based materials. Conventional biochar is highly sustainable, especially when produced from waste biomass with energy recovery. In summary, magnetic biochar provides an optimal balance across cost, recovery, and environmental impact. It is especially attractive for scalable, eco-friendly wastewater treatment applications where efficient separation, reusability, and affordability are critical. By combining the best qualities of carbon-based materials with the functionality of magnetic systems, magnetic biochar emerges as a superior choice in the evolving landscape of sustainable adsorbent

3.3.7 Recovery and regeneration

Magnetic biochar has important advantages in terms of recovery and regeneration, making it a practical and sustainable adsorbent for water treatment applications. One of the most attractive features of magnetic biochar is rapid and efficient separation from treated water using an external magnet, eliminating the need for energy-intensive filtration or centrifugation methods (Grzegorzek *et al.*, 2023). Once recovered, the saturated magnetic biochar can be regenerated using various desorption techniques such as acid washing (e.g. with HCl or HNO_3), alkaline treatment (e.g. NaOH), or salt solutions, which help remove adsorbed metal ions and nutrients from its surface. These methods are relatively simple and cost-effective, and their selection depends on the type of adsorbed contaminants. Studies have shown that magnetic biochar maintains good reusability performance over multiple adsorption–desorption cycles, typically sustaining 80–90% of its initial adsorption capacity even after

4–5 cycles, which underscores its economic viability. The regeneration process does not significantly alter its structural integrity or magnetic properties if done under controlled conditions. This recyclability not only reduces the cost associated with adsorbent replacement but also minimizes secondary waste generation. From an economic and environmental point of view, the reusability of magnetic biochar translates into reduced operational costs, lower raw material usage, and a smaller carbon footprint compared to single-use adsorbents such as activated carbon or non-magnetic biochar. Moreover, the possibility of regenerating and reusing magnetic biochar on site in decentralized treatment systems enhances its value in low-resource and rural areas.

3.4   Challenges and Future Prospects

3.4.1   Current limitations

Despite the promising performance of magnetic biochar in water treatment, several challenges hinder its widespread application and commercialization. One of the primary limitations is its stability under varying pH and temperature conditions. In highly acidic or alkaline environments, the structural reliability and surface functional groups of magnetic biochars may degrade, minimizing its adsorption efficiency. Additionally, extreme pH levels can affect the stability of entrenched magnetic particles such as Fe_3O_4, leading to partial dissolution or transformation into less active forms. Elevated temperatures may also affect the magnetic properties and surface reactivity, limiting the applicability of magnetic biochar in industrial effluents or thermal wastewater streams (Abbott et al., 2020). Another notable challenge is the relatively high production cost at large scales, particularly when high-purity iron salts or energy-intensive synthesis methods are used. While lab-scale preparation of magnetic biochar from agricultural biomass may be reasonable, scaling up the process to meet industrial demands without compromising performance or sustainability remains a hurdle.

3.4.2   Environmental concerns

Environmental concerns also arise, particularly related to iron leaching from MBC during repeated use or under acidic conditions, which can lead to secondary pollution in treated water. The release of iron ions may interfere with water chemistry and pose ecological risks, especially in sensitive aquatic environments. Furthermore, residual iron may complicate the downstream processing or reuse of treated water. To address these subjects and expand the utility of MBC, future research should focus on surface engineering strategies such as functionalizing biochar surfaces with polymers, metal oxides, or nanoparticles to improve pH resistance, enhance selectivity, and reduce leaching (Gęca et al., 2023). Incorporating nanocomposites or coatings can also improve mechanical stability and pollutant binding strength. Another promising avenue involves developing hybrid systems that integrate magnetic biochar with membrane technologies or AOPs. These create synergistic effects, improving overall treatment efficiency and enabling the simultaneous removal of multiple contaminants. Moreover, research should target the development of low-cost magnetic biochar from abundant waste biomass, such as invasive species like water hyacinth or municipal organic waste, thereby reducing raw material and production costs.

3.4.3 Circular economy approach

In the background of sustainability, adopting a circular economy approach is essential. One forward-thinking strategy involves resource recovery from spent magnetic biochar, such as retrieving adsorbed metals (e.g. Pb^{2+}, Cu^{2+}) for industrial reuse or converting saturated magnetic biochar into soil amendments or catalysts. This not only reduces waste but adds value to the spent material (Kacprzak *et al.*, 2022). Additionally, integrating magnetic biochar applications within circular frameworks can help close resource loops and minimize environmental footprints. In conclusion, while magnetic biochar holds enormous promise for water purification, overcoming technical and environmental challenges through innovative material design, hybrid treatment systems, and sustainable sourcing is key to its future success and contribution to circular, low-waste water management systems (Onukwulu *et al.*, 2022).

3.5 Conclusion

Magnetic biochar represents a significant advancement in the field of sustainable wastewater treatment. By combining the high adsorption efficiency of biochar with the easy recoverability of magnetic materials, it offers a multifunctional solution for the removal of diverse pollutants including dyes, heavy metals, pharmaceuticals, pesticides, and nutrients. Its enhanced properties such as increased surface area, porosity, and the presence of reactive functional groups enable superior contaminant capture through multiple mechanisms like electrostatic attraction, π–π interactions, and ion exchange. The magnetic nature of the material allows for efficient separation and reuse, addressing key operational and environmental challenges associated with traditional adsorbents.

This review underscores the material's broad applicability, high removal efficiency, and operational advantages, making it suitable for both laboratory and real-world wastewater treatment scenarios. While certain challenges such as cost-effective synthesis, pH sensitivity, and iron leaching remain, they can be addressed through ongoing research in material optimization, hybrid systems integration, and circular economy strategies. Overall, magnetic biochar holds strong potential as an eco-friendly, scalable, and economically viable technology for advanced water purification in both centralized and decentralized treatment systems.

References

Abbott, T., Kor-Bicakci, G., Islam, M.S. and Eskicioglu, C. (2020) A review on the fate of legacy and alternative antimicrobials and their metabolites during wastewater and sludge treatment. *International Journal of Molecular Sciences* 21(23), 9241. DOI: 10.3390/ijms21239241.

Akhtar, M.S., Ali, S. and Zaman, W. (2024) Innovative adsorbents for pollutant removal: Exploring the latest research and applications. *Molecules (Basel, Switzerland)* 29(18), 4317. DOI: 10.3390/molecules29184317.

Cheng, R., Hou, S., Wang, J., Zhu, H., Shutes, B. *et al.* (2022) Biochar-amended constructed wetlands for eutrophication control and microcystin (MC-LR) removal. *Chemosphere* 295, 133830. DOI: 10.1016/j.chemosphere.2022.133830.

Devi, M., Rawat, S. and Sharma, S. (2020) A comprehensive review of the pyrolysis process: From carbon nanomaterial synthesis to waste treatment. *Oxford Open Materials Science* 1(1), itab014. DOI: 10.1093/oxfmat/itab014.

Devi, R., Kumar, V., Kumar, S., Bulla, M., Jatrana, A. *et al.* (2023) Recent advancement in biomass-derived activated carbon for waste water treatment, energy storage, and gas purification: A review. *Journal of Materials Science* 58(30), 12119–12142. DOI: 10.1007/s10853-023-08773-0.

Dong, J., Shen, L., Shan, S., Liu, W., Qi, Z. *et al.* (2022) Optimizing magnetic functionalization conditions for efficient preparation of magnetic biochar and adsorption of Pb (II) from aqueous solution. *Science of The Total Environment* 806, 151442. DOI: 10.1016/j.scitotenv.2021.151442.

Gęca, M., Khalil, A.M., Tang, M., Bhakta, A.K., Snoussi, Y. *et al.* (2023) Surface treatment of biochar—methods, surface analysis and potential applications: A comprehensive review. *Surfaces* 6(2), 179–213. DOI: 10.3390/surfaces6020013.

Gillingham, M. (2022) Investigating a novel application for magnetic biochar: Practical barriers and policy considerations. Master's thesis, University of Nottingham, UK.

Gonçalves, J.O., Leones, A.R., Farias, B.S., Silva, M.D., Jaeschke, D.P. *et al.* (2025) A comprehensive review of agricultural residue-derived bioadsorbents for emerging contaminant removal. *Water* 17(14), 2141. DOI: 10.3390/w17142141.

Grzegorzek, M., Wartalska, K. and Kaźmierczak, B. (2023) Review of water treatment methods with a focus on energy consumption. *International Communications in Heat and Mass Transfer* 143, 106674. DOI: 10.1016/j.icheatmasstransfer.2023.106674.

Guarnotta, V., Amodei, R., Frasca, F., Aversa, A. and Giordano, C. (2022) Impact of chemical endocrine disruptors and hormone modulators on the endocrine system. *International Journal of Molecular Sciences* 23(10), 5710. DOI: 10.3390/ijms23105710.

Haider Jaffari, Z., Jeong, H., Shin, J., Kwak, J., Son, C. *et al.* (2023) Machine-learning-based prediction and optimization of emerging contaminants' adsorption capacity on biochar materials. *Chemical Engineering Journal* 466, 143073. DOI: 10.1016/j.cej.2023.143073.

Hu, B., Ai, Y., Jin, J., Hayat, T., Alsaedi, A. *et al.* (2020) Efficient elimination of organic and inorganic pollutants by biochar and biochar-based materials. *Biochar* 2(1), 47–64. DOI: 10.1007/s42773-020-00044-4.

Hu, Y., Guo, T., Ye, X., Li, Q., Guo, M. *et al.* (2013) Dye adsorption by resins: Effect of ionic strength on hydrophobic and electrostatic interactions. *Chemical Engineering Journal* 228, 392–397. DOI: 10.1016/j.cej.2013.04.116.

Ismail, M., Akhtar, K., Khan, M.I., Kamal, T., Khan, M.A. *et al.* (2019) Pollution, toxicity and carcinogenicity of organic dyes and their catalytic bio-remediation. *Current Pharmaceutical Design* 25(34), 3645–3663. DOI: 10.2174/138161282 5666191021142026.

Jacob, M.M., Ponnuchamy, M., Kapoor, A. and Sivaraman, P. (2024) Adsorptive membrane separation for eco-friendly decontamination of chlorpyrifos via biochar-impregnated cellulose acetate mixed matrix membrane. *Environmental Science and Pollution Research* 31, 56314–56331. DOI: 10.1007/ s11356-024-34912-4.

Kacprzak, M., Kupich, I., Jasinska, A. and Fijalkowski, K. (2022) Bio-based waste' substrates for degraded soil improvement—advantages and challenges in European context. *Energies* 15(1), 385. DOI: 10.3390/en15010385.

Kawamura, S. (2000) *Integrated Design and Operation of Water Treatment Facilities.* John Wiley & Sons.

Khan, M.Y., Mangrich, A.S., Schultz, J., Grasel, F.S., Mattoso, N. *et al.* (2015) Green chemistry preparation of superparamagnetic nanoparticles containing Fe_3O_4 cores in biochar. *Journal of Analytical and Applied Pyrolysis* 116, 42–48. DOI: 10.1016/j.jaap.2015.10.008.

Li, X., Xu, J., Luo, X. and Shi, J. (2022) Efficient adsorption of dyes from aqueous solution using a novel functionalized magnetic biochar: Synthesis, kinetics, isotherms, adsorption mechanism, and reusability. *Bioresource Technology* 360, 127526. DOI: 10.1016/j.biortech.2022.127526.

Liu, B., Vikrant, K., Kim, K.H., Kumar, V. and Kailasa, S.K. (2020) Critical role of water stability in metal–organic frameworks and advanced modification strategies for the extension of their applicability. *Environmental Science* 7(5), 1319–1347. DOI: 10.1039/C9EN01321K.

Long, M., Zhou, C., Xia, S. and Guadiea, A. (2017) Concomitant cr(VI) reduction and cr(III) precipitation with nitrate in a methane/oxygen-based membrane biofilm reactor. *Chemical Engineering Journal* 315, 58–66. DOI: 10.1016/j. cej.2017.01.018.

Lu, L., Yu, W., Wang, Y., Zhang, K., Zhu, X. *et al.* (2020) Application of biochar-based materials in environmental remediation: From multi-level structures to specific devices. *Biochar* 2(1), 1–31. DOI: 10.1007/s42773-020-00041-7.

Lyu, J., Zhu, L. and Burda, C. (2014) Considerations to improve adsorption and photocatalysis of low concentration air pollutants on TiO2. *Catalysis Today* 225, 24–33. DOI: 10.1016/j.cattod.2013.10.089.

Manyangadze, M., Chikuruwo, N.H.M., Narsaiah, T.B., Chakra, C.S., Radhakumari, M. *et al.* (2020) Enhancing adsorption capacity of nano-adsorbents via surface modification: A review. *South African Journal of Chemical Engineering* 31(1), 25–32. DOI: 10.1016/j.sajce.2019.11.003.

Mays, T.J. (2007) A new classification of pore sizes. In: *Studies in Surface Science and Catalysis*, Vol. 160. Elsevier, pp. 57–62. DOI: 10.1016/S0167-2991(07)80009-7.

Meng, H., Chen, Z., Wei, W., Xu, J., Duan, H. *et al.* (2025) Magnetic hydrochar for sustainable wastewater management. *Npj Materials Sustainability* 3(1), 7. DOI: 10.1038/s44296-024-00047-3.

Nikić, J., Watson, M., Tubić, A., Šolić, M. and Agbaba, J. (2024) Recent trends in the application of magnetic nanocomposites for heavy metals removal from water: A review. *Separation Science and Technology* 59(2), 293–331. DOI: 10.1080/01496395.2024.2315626.

Niu, H., Shi, S., Zhu, S., Cai, Y. and Cao, D. (2025) Biochars-inlaided nano zero-valent iron reactors: A tool for visualized analysis of soil-nanomaterials micro-interfacial interaction in soil remediation process. *The Science of the Total Environment* 958, 177829. DOI: 10.1016/j.scitotenv.2024.177829.

Onukwulu, E.C., Agho, M.O. and Eyo-Udo, N.L. (2022) Circular economy models for sustainable resource management in energy supply chains. *World Journal of Advanced Science and Technology* 2(2), 034–057. DOI: 10.53346/wjast.2022.2.2.0048.

Psarommatis, F. and May, G. (2025) A cost–benefit model for sustainable product reuse and repurposing in circular remanufacturing. *Sustainability* 17(1), 245. DOI: 10.3390/su17010245.

Ramos Guivar, J.A., Sadrollahi, E., Menzel, D., Ramos Fernandes, E.G., López, E.O. *et al.* (2017) Magnetic, structural and surface properties of functionalized maghemite nanoparticles for copper and lead adsorption. *RSC Advances* 7(46), 28763–28779. DOI: 10.1039/C7RA02750H.

Rezania, S., Ponraj, M., Talaiekhozani, A., Mohamad, S.E., Md Din, M.F. *et al.* (2015) Perspectives of phytoremediation using water hyacinth for removal of heavy metals, organic and inorganic pollutants in wastewater. *Journal of Environmental Management* 163, 125–133. DOI: 10.1016/j.jenvman.2015.08.018.

Riaz, U., Rafi, F., Naveed, M., Mehdi, S.M., Murtaza, G. *et al.* (2021) Pesticide pollution in an aquatic environment. In: *Freshwater Pollution and Aquatic Ecosystems*. Apple Academic Press, pp. 131–163.

Roper, M. and Seminara, A. (2019) Mycofluidics: The fluid mechanics of fungal adaptation. *Annual Review of Fluid Mechanics* 51(1), 511–538. DOI: 10.1146/annurev-fluid-122316-045308.

Shakoor, M.B., Ali, S., Rizwan, M., Abbas, F., Bibi, I. *et al.* (2020) A review of biochar-based sorbents for separation of heavy metals from water. *International Journal of Phytoremediation* 22(2), 111–126. DOI: 10.1080/15226514.2019.1647405.

Sharma, G. and Kamalesu, S. (2023) Review on sustainable synthesis of semi-amorphous ti-BDS MOF material, activated carbon, and graphene. *Materials Today: Proceedings*. DOI: 10.1016/j.matpr.2023.01.249.

Siddiqui, A.J., Kumari, N., Adnan, M., Kumar, S., Abdelgadir, A. *et al.* (2023) Impregnation of modified magnetic nanoparticles on low-cost agro-waste-derived biochar for enhanced removal of pharmaceutically active compounds: Performance evaluation and optimization using response surface methodology. *Water* 15(9), 1688. DOI: 10.3390/w15091688.

Smidt, E. and Lechner, P. (2005) Study on the degradation and stabilization of organic matter in waste by means of thermal analyses. *Thermochimica Acta* 438(1–2), 22–28. DOI: 10.1016/j.tca.2005.08.013.

Srai, J.S., Kumar, M., Graham, G., Phillips, W., Tooze, J. *et al.* (2016) Distributed manufacturing: Scope, challenges and opportunities. *International Journal of Production Research* 54(23), 6917–6935. DOI: 10.1080/00207543.2016.1192302.

Tang, R., Gong, D., Deng, Y., Xiong, S., Zheng, J. *et al.* (2022) π-π stacking derived from graphene-like biochar/g-C_3N_4 with tunable band structure for photocatalytic antibiotics degradation via peroxymonosulfate activation. *Journal of Hazardous Materials* 423(Pt A), 126944. DOI: 10.1016/j.jhazmat.2021.126944.

Tartaj, P. and Amarilla, J.M. (2011) Iron oxide porous nanorods with different textural properties and surface composition: Preparation, characterization and electrochemical lithium storage capabilities. *Journal of Power Sources* 196(4), 2164–2170. DOI: 10.1016/j.jpowsour.2010.09.116.

Tomei, M.C., Mosca Angelucci, D., Mascolo, G. and Kunkel, U. (2019) Post-aerobic treatment to enhance the removal of conventional and emerging micropollutants in the digestion of waste sludge. *Waste Management (New York, N.Y.)* 96, 36–46. DOI: 10.1016/j.wasman.2019.07.013.

Velusamy, K., Periyasamy, S., Kumar, P.S., Vo, D.-V.N., Sindhu, J. *et al.* (2021) Advanced techniques to remove phosphates and nitrates from waters: A review. *Environmental Chemistry Letters* 19(4), 3165–3180. DOI: 10.1007/s10311-021-01239-2.

Xia, G. and Ball, W.P. (1999) Adsorption-partitioning uptake of nine low-polarity organic chemicals on a natural sorbent. *Environmental Science & Technology* 33(2), 262–269. DOI: 10.1021/es980581g.

Zeng, X., Zhang, G., Zhu, J. and Wu, Z. (2022) Adsorption of heavy metal ions in water by surface functionalized magnetic composites: A review. *Environmental Science: Water Research & Technology* 8(5), 907–925. DOI: 10.1039/D1EW00868D.

Zinicovscaia, I. (2016) *Conventional Methods of Wastewater Treatment.* Cyanobacteria for Bioremediation of Wastewaters, pp. 17–25.

Adsorption Isotherm and Kinetic Modelling for Magnetic Biochar

4

4.1 Introduction

For scaling any water treatment plant, it is essential to understand and optimize the process, especially for the treatment of dye and heavy metal from materials like activated carbon, biochar, or magnetic bioadsorbent. A quantitative basis for how contaminants interact with adsorbents under a variety of environmental and operational conditions is needed, where a systematic approach is applied rather than trusting empirical trial-and-error methods. So, the importance of modelling, such as determining the adsorption capacity, surface affinity, reaction kinetics, and energy changes, is essential in the effective design of a water treatment system. Among the key tools in this process are adsorption isotherms and kinetic models, which are used to explaining the adsorption mechanism. The isotherms provide the understanding of equilibrium distribution of adsorbate between the liquid and solid phases. For example, the Langmuir isotherm considers monolayer adsorption on a homogeneous surface with finite sites, signifying chemisorption, whereas the Freundlich model is more suitable for heterogeneous surfaces and multilayer adsorption, representing the physisorption (Kumar *et al.*, 2019). Other models such as the Temkin isotherm aid in representing the adsorption interaction between the adsorbate–adsorbate repulsion or interaction across the adsorbent surface. These models use different parameters that collectively help to appraise the favourability and mechanism of the adsorption.

Kinetic models explain the rate and pathway of adsorption, controlling the design of reactors and determining contact time, output, and regeneration frequency (Mohammed *et al.*, 2011). The pseudo-first-order (PFO) model, in which the rate is proportional to unoccupied sites, typically represents physical

Corresponding author: danbahadur.chem@gmail.com

© CAB International 2026. *Magnetic Biochar for Wastewater Remediation in the Textile Industry* (D. Bahadur Pal and A. Kapoor)

adsorption, whereas the pseudo-second-order (PSO) model considers chemisorption involving electron exchange. Additionally, intra-particle diffusion models are used to identify whether diffusion within pores is the rate-limiting step, which is common among porous adsorbents such as biochar or nano composites (Mohamed Nasser *et al.*, 2024).

Together, isotherms and kinetics deliver a complete picture of the adsorption mechanism and performance under different operational setups, which is crucial in moving from lab-scale experiments to pilot or industrial-scale applications. This becomes especially relevant in the treatment of wastewater containing complex mixtures of dyes (e.g. methylene blue, Congo red,) or heavy metals (e.g. Cr^{6+}, Pb^{2+}, Cd^{2+}), where adsorption performance is influenced by multiple variables such as pH, temperature, ionic strength, adsorbent surface chemistry, and pollutant concentration (Aldegs *et al.*, 2008). Modelling considers the selection and optimization of these parameters, thereby improving removal efficiency and cost-effectiveness. Furthermore, in designing large-scale treatment systems such as fixed-bed columns or fluidized reactors, accurate adsorption models help predict breakthrough curves, adsorption zone lengths, regeneration cycles, and bed life. For example, in textile effluent treatment, kinetic and isotherm modelling allows the determination of the optimal adsorbent dosage, contact time, and flow rate to maximize dye removal and minimize operational cost. Similarly, for metal removal, modelling helps identify adsorbents with selective affinity, evaluate competitive adsorption in multi-metal systems, and prevent early saturation of the adsorbent bed (Mahamadi, 2019). Moreover, understanding the kinetic control mechanisms, whether surface reaction or intra-particle diffusion, enables proper reactor configuration and reduces energy or material input. Additionally, adsorption modelling assists in life cycle and sustainability assessments of the adsorbent, including regeneration potential and post-use disposal, which are essential for environmental compliance and economic feasibility.

4.2   Adsorption Isotherm Models

Adsorption isotherm models are important for understanding and predicting how adsorbate molecules interact with adsorbent surfaces at equilibrium. These models deliver a mathematical association between the amounts of adsorbate adsorbed per unit mass of adsorbent and the equilibrium concentration of adsorbate in the solution. The primary aim of adsorption isotherms is to explain the adsorption mechanism, to calculate the adsorption capacity, predict system behaviour, design large-scale treatment processes, and compare the performance of numerous adsorbents (Karunakaran *et al.*, 2022). By fitting experimental data to isotherm equations, researchers and engineers can determine whether the adsorption process involves monolayer or multilayer formation, whether it occurs on a homogeneous or heterogeneous surface, and whether it is dominated by physical or chemical forces. At equilibrium, adsorption and desorption rates balance, and the system reaches a steady state. The contact of adsorbate with the adsorbent depends on the surface characteristics of the adsorbent, including surface area, porosity, and the presence of active functional groups such as hydroxyl, carboxyl, or amine groups, which enhance binding.

In physical adsorption (physisorption), weak van der Waals forces are involved, and the process is usually rapid, reversible, and may lead to multi-layer adsorption, especially at lower temperatures (Nielaba *et al.*, 1994). In contrast, chemical adsorption (chemisorption) involves stronger covalent or ionic bonds, occurs more slowly, forms only a monolayer, and is often irreversible and selective for specific adsorbate molecules. Common isotherm models include: the Langmuir isotherm, which assumes uniform adsorption onto a surface with finite and identical sites leading to monolayer formation; the Freundlich isotherm, which describes adsorption on heterogeneous surfaces and allows for multilayer adsorption; and models like Temkin that help understand adsorption energy distribution and distinguish between physical and chemical adsorption mechanisms (Aljamali *et al.*, 2021). These models not only aid in interpreting laboratory results but are also essential for scaling up adsorption systems for real-world applications such as dye and heavy metal removal from wastewater. They allow an estimation of the adsorbent quantity required, efficiency of pollutant removal, and optimization of operating conditions. Adsorption isotherm models are very important tools in environmental engineering, material science, and chemical process industries. They are provide both theoretical and practical approaches to determining the interaction of adsorbent and adsorbate at equilibrium. A key factor is measuring the adsorption capacity at equilibrium (q_e), which quantifies the amount of biomass adsorbed per unit mass of the bioadsorbent (Bruch *et al.*, 2007). Furthermore, the amount of biomass adsorbed per unit mass of the adsorbent at equilibrium can be calculated as:

$$q_e = \frac{V(c_o - c_e)}{m} \tag{1}$$

$$q_t = \frac{V(c_o - c_t)}{m} \tag{2}$$

Where c_o, c_e, and c_t signify the concentration at initial, equilibrium, and time t stages, respectively; m is the weight or mass of the adsorbent (g); V is the volume of the solution (l); And q_e and q_t represent the adsorption capacity (mg/g) at equilibrium and at any given time t, respectively. The removal efficiency (E) of the adsorbent can be calculated as:

$$E(\%) = \frac{(c_o - c_e)}{c_o} * 100 \tag{3}$$

4.2.1 Langmuir adsorption isotherm

The Langmuir model or concept predicts whether the adsorbed molecules form a monolayer, or unimolecular layer, on the surface of the adsorbent by chemical reaction. This model is effective for describing adsorption driven by specific chemical interactions between the adsorbate and the adsorbent surface (Weber and Smith, 1987). It assumes that all adsorption sites are equivalent and that once a dye molecule occupies a site, no further adsorption can occur at that specific location. The linear form of the Langmuir model is commonly expressed as:

$$\frac{c_t}{q_e} = \frac{1}{q_m\,_{K_L}} + \frac{c_e}{q_m} \tag{4}$$

Where q_m (mg/g) is the maximum amount of dye adsorbed within a certain interval, and the Langmuir constant is K_L (l/mg). A significant characteristic R_L is defined as:

$$R_L = \frac{1}{\left(1 + c_0\,K_L\right)} \tag{5}$$

The value of R_L indicates the type of isotherm: irreversible ($R_L = 0$), unfavourable ($R_L > 1$), linear ($R_L = 1$), or favourable ($0 < R_L < 1$), However, the Langmuir isotherm has limitations. It does not account for multilayer adsorption, is insufficient for heterogeneous surfaces with varying energy sites, and assumes no lateral interaction between adsorbed molecules. Furthermore, its accuracy decreases at high adsorbate concentrations where real systems may stray from ideal monolayer behaviour. Despite these limitations, the Langmuir model remains a basic adsorption science for system design and material selection due to its clarity and broad applicability.

4.2.2 Freundlich adsorption isotherm

The Freundlich isotherm is an empirical model widely used to describe adsorption on heterogeneous surfaces, especially when multilayer adsorption occurs. Unlike the Langmuir isotherm's assumption of a uniform surface and monolayer adsorption, the Freundlich model is based on heterogeneous surface, featuring a non-uniform distribution of adsorption heats and affinities (Tompkins, 1950). It also accommodates multilayer adsorption, making it more suitable for real-world systems such as activated carbon or biochar-based adsorbents. The Freundlich adsorption isotherm, which describes adsorption on heterogeneous surfaces with varying affinities, offers a valuable approach to the multilayer adsorption behaviour dye on bioadsorbent (Mussa *et al.*, 2023). The Freundlich model posits that adsorption capacity increases with rising adsorbate concentration, reflecting a distribution of adsorption site energies, for which the linear equation is given by:

$$\ln q_e = \ln K_F + \frac{1}{n}\ln c_e \tag{6}$$

Where q_e is the amount of solute adsorbed per unit weight of adsorbent (mg/g), and c_e is the equilibrium concentration of solute in solution (mg/l). The constant K_F represents the adsorption capacity, while $1/n$ represents the adsorption intensity and surface heterogeneity. Values of $1/n$ between 0 and 1 suggest better adsorption, with lower values indicating stronger adsorbate–adsorbent interactions. The Freundlich isotherm does not predict a saturation point, reducing its reliability at higher concentrations.

4.2.3   Temkin isotherm

The Temkin isotherm is very important model that is employed to designate adsorption processes and is characterized by contacts between the adsorbent and adsorbate, where the heat of adsorption linearly decreases with increasing surface treatment. Unlike the Langmuir model's assumption of a constant heat of adsorption, the Temkin isotherm acknowledges that, in real adsorption systems, the adsorption energy reduces as more sites become occupied (Travin et al., 2019). This model considers the interactions between adsorbate molecules on the adsorbent surface, leading to a uniform distribution of binding energies. The Temkin model presents a linear reduction in adsorption energy which is represented by:

$$q_e = B\ln(A\,c_e) \tag{7}$$

Where B is heat of absorption expressed as $B = R_T/b$, in which R is the gas constant, and T is temperature, while A is the Temkin constant related to absorption capacity. The Temkin model has been widely studied in dye adsorption, heavy metal removal, and pollutant uptake by various materials. This model is especially valuable when adsorbate–adsorbate interactions are significant and cannot be ignored, such as in systems with polar molecules or ions. However, it is best suited for intermediate concentration ranges and may not fit well at very low or high concentrations. The Temkin isotherm provides insights into the thermodynamics of adsorption, indicating how adsorption energy changes and whether the process is endothermic or exothermic.

4.2.4   Dubinin–Radushkevich (D–R) isotherm

The D–R isotherm is commonly applied to describe adsorption processes occurring on microporous and heterogeneous surfaces (Dubinin, 1989). Unlike the Langmuir model, which assumes uniform surface energy, the D–R model allows calculation of the mean free energy of adsorption, which is often used to determine the nature of adsorption, whether it is physical or chemical. The model is represented by:

$$q_e = q_m\exp(-B\varepsilon^2) \tag{8}$$

Where $\varepsilon = R_1 \ln (1 + 1/c_e)$ represents the Polanyi potential; $-q_e$ is the amount adsorbed at equilibrium (mg/g); and q_m is the amount adsorbed at equilibrium at time t (mg/g). If the value of the mean free energy (E) of adsorption is $<8\,kJ/mol$, this indicates physical adsorption, whereas values between 8 and $16\,kJ/mol$ suggest chemical ion-exchange interactions.

4.2.5   Harkins–Jura isotherm

This isotherm provides an insight into multilayer adsorption and the heterogeneous nature of adsorbent surfaces (Nandiyanto *et al.*, 2024). The model is represented by the equation:

$$\frac{1}{qe^2} = \frac{B}{A} - \frac{1}{A}\log c_e \tag{9}$$

Where A and B are Harkins and Jura constants related to adsorption capacity. This isotherm model is particularly suitable for adsorbents with varied surface energies, such as those derived from agricultural wastes or natural minerals. In practice, this model has been used to describe Congo red adsorption on modified biochar, where a good fit to the Harkins–Jura model suggests multilayer adsorption due to electrostatic attraction and π-π interactions between dye molecules.

4.3   Adsorption Kinetics

Adsorption kinetics represents how fast molecules adsorb to a surface. It is important for all stages of cleaning water, where the molecules first move to the surface, then diffuse into the pores of the material, and finally attach to active sites. Understanding which of these steps is rate-limiting and slowest helps to improve the process (Murdoch, 1981). Factors such as the type of molecule, the surface it adheres to, temperature, and concentration all play a role in how fast adsorption proceeds. Recent research focuses on developing better models and materials for more efficient adsorption in various applications. This information is vital for optimizing dye removal because it provides insights into how quickly adsorbate molecules are captured by the adsorbent's surface over time.

4.3.1   Pseudo-first-order (PFO) kinetics

The PFO kinetic model is based on the adsorption rate being directly proportional to the number of available adsorption sites on the adsorbent's surface (Revellame *et al.*, 2020). The expression can be represented as:

$$\ln(q_e - q_t) = \ln(q_e) - \frac{K_1}{2.303}*\tag{10}$$

The model is widely employed in adsorption studies to characterize the rate of adsorption as being proportional to the number of unoccupied sites on the adsorbent surface, assuming that physisorption is the leading mechanism. This model facilitates the estimation of the rate constant and the forecast of the time required to reach equilibrium, thereby aiding in the design and optimization of both batch and fixed-bed adsorption systems (Fischer and Freund, 2020). The assumption that physisorption is the primary interaction mechanism provides a basis for this model, making it useful for comparing different adsorbents or operating conditions. It is often utilized in combination with more compound models to gain a comprehensive understanding of adsorption behaviour. The model's ease of application and relevance in initial kinetic screening established it as a fundamental tool in environmental and industrial adsorption research.

4.3.2 Pseudo-second order (PSO) kinetics

The PSO model is used when the adsorption process involves chemisorption, where the rate of adsorption is directly proportional to the square of the adsorbate concentration on the surface (Lombardo and Bell, 1991). The equation is expressed as:

$$\frac{t}{q_t} = \frac{1}{(K_2 * q_e 2)} + \frac{1}{qe} * t \tag{11}$$

In this equation, t represents time (min), q_t is the adsorption capacity at time t (mg/g), q_e is the adsorption capacity at equilibrium (mg/g), and K_2 is the PSO rate constant (mg/g·min). The higher value of K_2 for Methylene Blue (MB) indicates a faster adsorption process, highlighting strong chemisorptive interactions between the cationic MB dye molecules and the negatively charged functional groups on the bioadsorbent surface. This model has significant applications in the removal of dyes and heavy metals from wastewater using materials such as activated carbon, agricultural biochar, magnetic biochar, and nanocomposites (Moosavi et al., 2020). It is particularly preferred for systems where chemical bonding, rather than solely physical interactions, dictates the adsorption process. The model is useful when adsorption capacity is expected and in designing scale-up systems like packed bed reactors. A good fit indicates chemisorption, offering insights for selecting or modifying adsorbents to enhance performance. The PSO model is suitable for engineering applications, and efficient, and sustainable adsorption systems for environmental remediation.

4.3.3 Intra-particle diffusion

The intra-particle diffusion model is useful for the mechanism of adsorption, specifically how dye molecules permeate within the pores of the bioadsorbent

(Tran *et al.*, 2017). This model is important for determining whether the adsorption process is primarily measured by intra-particle diffusion or influenced by other stages, such as external mass transfer. The model can be expressed as:

$$q_t = K_{id} \, t^{0.5} + c \tag{12}$$

The slope of the resulting line indicates the rate constant, while the intercept provides insight into the boundary layer thickness. Here, q_t, the amount of adsorbate adsorbed at time t (mg/g), K_{id} is the intra-particle diffusion rate constant, t is time (min), and c is a constant related to the boundary layer thickness. Applications of this model are prevalent in wastewater treatment systems for the removal of dyes, heavy metals, and organic pollutants, particularly when adsorbents possess high porosity and large surface areas (Kumar *et al.*, 2025). It helps in identifying whether external surface adsorption, intraparticle diffusion, or both control the rate of adsorption. The significance of this model lies in its ability to diagnose the complexity of the adsorption process and guide the design of more effective adsorbents and contact times (Weber and Smith, 1987). It is essential for optimizing system performance in batch and column reactors, especially where quick and deep diffusion into the adsorbent is required for efficient pollutant capture. Table 4.1 summarizes the adsorption kinetic models.

4.4   Conclusion

Kinetic and isotherm modelling play important roles in determining the mechanisms of adsorption processes in wastewater treatment. By applying isotherm models, researchers can estimate the adsorption, whether it occurs on homogeneous or heterogeneous surfaces, and in monolayer or multilayer. These models help to estimate essential parameters such as maximum adsorption capacity and adsorption intensity, which are important for estimating the efficiency of adsorbents such as magnetic biochar. Similarly, kinetic models such provide welcome insights into the rate and mechanisms of adsorption, whether governed by surface reactions or diffusion processes. Among these, the PSO model frequently demonstrates the best fit for magnetic biochar, indicating chemisorption as the dominant mechanism. The effective use of these models not only supports an understanding of experimental data but also aids in optimizing process conditions for large-scale applications. Tools such as Origin and Excel enhance model fitting, and isotherm and kinetic modelling are important tools for designing efficient, cost-effective, and scalable wastewater treatment systems using magnetic biochar. These models deliver a scientific foundation for optimizing adsorption performance, contributing significantly to the development of sustainable environmental technologies.

Table 4.1. Adsorption kinetic models.

Model	Assumptions / basis	Linear equation	Parameters	Plot for linearity	Indication
Pseudo-first-order (PFO)	The rate of adsorption is proportional to the number of unoccupied sites	$\log(q_e - q_t) = \log(q_e) - (K_1/2.303)t$	q_e = equilibrium adsorption capacity (mg/g) K_1 = rate constant (1/min)	$\log(q_e - q_t)$ vs t	Suitable for initial adsorption stages
Pseudo-second-order (PSO)	Adsorption involves chemisorption (electron sharing/exchange between adsorbent/adsorbate)	$t/q_t = 1/(K_2 \cdot q_e^2) + t/q_e$	q_E = equilibrium adsorption capacity (mg/g) K_2 = rate constant (g/mg·min)	t/q_t vs t	Best fit when adsorption is chemical in nature
Intraparticle diffusion	Adsorption rate is controlled by diffusion within particles	$q_t = K_{id} \cdot t^{0.5} + c$	K_{id} = intraparticle diffusion rate constant c = intercept related to boundary layer thickness	q_t vs $t^{0.5}$	Multi-linear curve suggests multiple mechanisms; if line passes through origin, intraparticle diffusion is sole rate-limiting step

References

Aldegs, Y., Elbarghouthi, M., Elsheikh, A. and Walker, G. (2008) Effect of solution pH, ionic strength, and temperature on adsorption behavior of reactive dyes on activated carbon. *Dyes and Pigments* 77(1), 16–23. DOI: 10.1016/j. dyepig.2007.03.001.

Aljamali, N.M., Khdur, R. and Alfatlawi, I.O. (2021) Physical and chemical adsorption and its applications. *International Journal of Thermodynamics and Chemical Kinetics* 7(2), 1–8.

Bruch, L.W., Cole, M.W. and Zaremba, E. (2007) *Physical Adsorption: Forces and Phenomena.* Courier Dover Publications.

Dubinin, M.M. (1989) Fundamentals of the theory of adsorption in micropores of carbon adsorbents: Characteristics of their adsorption properties and microporous structures. *Carbon* 27(3), 457–467. DOI: 10.1016/0008-6223(89)90078-X.

Fischer, K.L. and Freund, H. (2020) On the optimal design of load flexible fixed bed reactors: Integration of dynamics into the design problem. *Chemical Engineering Journal* 393, 124722. DOI: 10.1016/j.cej.2020.124722.

Karunakaran, K., Usman, M. and Sillanpää, M. (2022) A review on superadsorbents with adsorption capacity ≥ 1000 mg g^{-1} and perspectives on their upscaling for water/wastewater treatment. *Sustainability* 14(24), 16927. DOI: 10.3390/su142416927.

Kumar, A., Kapoor, A., Kumar Rathoure, A., Lal Devnani, G. and Pal, B.D. (2025) Organic compounds removal using magnetic biochar from textile industries based wastewater–a comprehensive review. *Sustainable Processes Connect* 1, 0012. DOI: 10.69709/SusProc.2025.147565.

Kumar, K.V., Gadipelli, S., Wood, B., Ramisetty, K.A., Stewart, A.A. *et al.* (2019) Characterization of the adsorption site energies and heterogeneous surfaces of porous materials. *Journal of Materials Chemistry A* 7(17), 10104–10137. DOI: 10.1039/C9TA00287A.

Lombardo, S.J. and Bell, A.T. (1991) A review of theoretical models of adsorption, diffusion, desorption, and reaction of gases on metal surfaces. *Surface Science Reports* 13(1–2), 3–72. DOI: 10.1016/0167-5729(91)90004-H.

Mahamadi, C. (2019) On the dominance of Pb during competitive biosorption from multi-metal systems: A review. *Cogent Environmental Science* 5(1), 1635335. DOI: 10.1080/23311843.2019.1635335.

Mohamed Nasser, S., Abbas, M. and Trari, M. (2024) Understanding the rate-limiting step adsorption kinetics onto biomaterials for mechanism adsorption control. *Progress in Reaction Kinetics and Mechanism* 49, 14686783241226858. DOI: 10.1177/14686783241226858.

Mohammed, F.M., Roberts, E.P.L., Hill, A., Campen, A.K. and Brown, N.W. (2011) Continuous water treatment by adsorption and electrochemical regeneration. *Water Research* 45(10), 3065–3074. DOI: 10.1016/j.watres.2011.03.023.

Moosavi, S., Lai, C.W., Gan, S., Zamiri, G., Akbarzadeh Pivehzhani, O. *et al.* (2020) Application of efficient magnetic particles and activated carbon for dye removal from wastewater. *ACS Omega* 5(33), 20684–20697.

Murdoch, J.R. (1981) What is the rate-limiting step of a multistep reaction? *Journal of Chemical Education* 58(1), 32.

Mussa, Z.H., Al-Ameer, L.R., Al-Qaim, F.F., Deyab, I.F., Kamyab, H. *et al.* (2023) A comprehensive review on adsorption of methylene blue dye using leaf waste as a bio-sorbent: Isotherm adsorption, kinetics, and thermodynamics studies. *Environmental Monitoring and Assessment* 195(8), 940. DOI: 10.1007/s10661-023-11432-1.

Nandiyanto, A.B.D., Putri, M.E., Fiandini, M., Ragadhita, R., Kurniawan, T. *et al.* (2024) Characteristics of ammonia adsorption on various sizes of calcium carbonate microparticles from chicken eggshell waste. *Moroccan Journal of Chemistry* 12(3), 1073–1096. DOI: 10.48317/IMIST.PRSM/morjchem-v12i3.48106.

Nielaba, P., Privman, V. and Wang, J.-S (1994) Irreversible multilayer adsorption. *Berichte Der Bunsengesellschaft Für Physikalische Chemie* 98(3), 451–454. DOI: 10.1002/bbpc.19940980340.

Revellame, E.D., Fortela, D.L., Sharp, W., Hernandez, R. and Zappi, M.E. (2020) Adsorption kinetic modeling using pseudo-first order and pseudo-second order rate laws: A review. *Cleaner Engineering and Technology* 1, 100032. DOI: 10.1016/j.clet.2020.100032.

Tompkins, F.C. (1950) Adsorption isotherms for non-uniform surfaces. *Transactions of the Faraday Society* 46, 580. DOI: 10.1039/tf9504600580.

Tran, H.N., You, S.J., Nguyen, T.V. and Chao, H.P. (2017) Insight into the adsorption mechanism of cationic dye onto biosorbents derived from agricultural wastes. *Chemical Engineering Communications* 204(9), 1020–1036. DOI: 10.1080/00986445.2017.1336090.

Travin, S.O., Gromov, O.B., Utrobin, D.V. and Roshchin, A.V. (2019) Kinetic simulation of adsorption isotherms. *Russian Journal of Physical Chemistry B* 13(6), 975–985. DOI: 10.1134/S1990793119060113.

Weber, W.J. and Smith, E.H. (1987) Simulation and design models for adsorption processes. *Environmental Science & Technology* 21(11), 1040–1050. DOI: 10.1021/es00164a002.

Mechanisms of Pollutant Removal Using Magnetic Biochar

5

5.1 Introduction

The concern over environmental pollution, particularly in water ecosystems, has extended the demand for efficient, sustainable, and cost-effective treatment methods. Of the techniques being developed and optimized, adsorption stands out as one of the most effective, owing to its simplicity, high efficiency, and adaptability to a wide range of pollutants (Akhtar *et al.*, 2024). The use of magnetic biochar as an advanced adsorbent material within this approach has transformed wastewater treatment methodology. To optimize the use of magnetic bichar and design better effective adsorption systems as treatment strategies, however, it is essential to thoroughly understand the mechanisms that govern pollutant removal. Removal mechanisms determine how a pollutant interacts with an adsorbent, whether as physical forces like van der Waals interactions, chemical bonding, electrostatic attraction, ion exchange, or surface precipitation (Takio *et al.*, 2021). A complete understanding of these mechanisms not only helps in estimating the adsorption behaviour under various environmental conditions such as pH, temperature, and ionic strength, but also helps in the regeneration or recycling of the adsorbent. Moreover, knowledge of removal mechanisms allows the fine-tuning of material properties to target specific pollutants, enhancing selectivity and adsorption capacity (Dutta *et al.*, 2024). For example, certain functional groups on the adsorbent surface may specially interact with heavy metals, while others may be more active for binding organic dyes. Thus, understanding these mechanisms is essential for achieving high removal efficiency, minimizing secondary pollution, and guiding the development of next-generation adsorbents.

Corresponding author: danbahadur.chem@gmail.com

© CAB International 2026. *Magnetic Biochar for Wastewater Remediation in the Textile Industry* (D. Bahadur Pal and A. Kapoor)

The link between pollutant type and removal pathway is important. Pollutants can broadly be classified into organic, as pesticides and pharmaceuticals dyes, and inorganic, as heavy metals, nitrate, fluoride types, each of which follows different adsorption mechanisms or pathways. Organic pollutants often undergo removal through hydrogen bonding, and hydrophobic effects, and $\pi-\pi$ interactions, particularly when aromatic structures are involved. In contrast, inorganic pollutants such as metal ions typically bind through ion exchange, electrostatic attraction, and functional groups such as hydroxyl, carboxyl, or amine groups (Björneholm *et al.*, 2022). Identifying this difference in mechanism is essential for the design of biochar materials with custom-made surface chemistries. For example, nitrogen fixing may enhance attraction for heavy metals by introducing amine or imine groups, whereas carbonization temperature can influence pore-size distribution to better arrest organic molecules. Therefore, the pollutant-specific understanding of removal pathways facilitates the engineering of adsorbents, increasing their overall efficacy in different treatment setups.

Magnetic biochar, a modified form of conventional biochar, integrates iron-based magnetic particles like Fe_3O into the carbon matrix, significantly improving the structural and functional properties (Shaheen *et al.*, 2022). The structure and surface chemistry of magnetic biochar play an essential role in improving pollutant removal. The magnetic nature permits the easy separation of the adsorbent from the treated water using external magnets, which makes the process more efficient and sustainable. The porous nature of biochar provides a large specific surface area, allowing wide-ranging pollutant interaction. Furthermore, the incorporation of iron oxides introduces additional activity, which is useful for the removal of heavy metals such as arsenic, lead, and chromium. Surface modifications can also introduce oxygen-containing functional groups like hydroxyl, carboxyl group or nitrogen functionalities, which improve the polarity and interactions with both hydrophilic and hydrophobic contaminants (Wang *et al.*, 2007). Additionally, magnetic biochar can enhance the degradation of certain organic pollutants when activated by hydrogen peroxide, further extending its utility beyond simple adsorption. In summary the integration of removal mechanism understanding, pollutant-specific adsorption pathways, and advanced material engineering such as magnetic biochar synthesis presents a comprehensive strategy for addressing modern water pollution challenges. As global freshwater resources face increasing stress, developing such intelligent and multifunctional materials will be key to ensuring safe and sustainable water management (Nagal and Prabhakar, 2025).

5.2   Different Adsorption Mechanisms

Adsorption is a surface phenomenon in which atoms, ions, or molecules from a gas, liquid, or dissolved solid collect on the surface of a solid or liquid adsorbent. Unlike absorption, where a substance diffuses into the bulk of a material, adsorption is restricted to the surface and forming a thin layer of adsorbate (Alaqarbeh, 2021). This process is mostly used in water purification, air filtration, catalysis, and sensor technology due to its simplicity and high efficiency. The interface between the adsorbate and the adsorbent can be classified into two main types: physical adsorption (physisorption) and chemical adsorption (chemisorption).

5.2.1   Physical adsorption

Physical adsorption involves the build-up of adsorbate molecules on a surface through weak van der Waals forces, which contain dipole–dipole interactions, dispersion forces, and hydrogen bonding. These forces generally contain minimum energy of less than 40 kJ/mole, which does not include any chemical bond formation or electron exchange (Compaan *et al.*, 2008). Physisorption occurs at low to moderate temperatures, does not require high activation energy, and is reversible in nature. It is characterized by multilayer adsorption, which makes it ideal for systems in which large surface areas are available, such as in porous materials like activated carbon or biochar (Lu *et al.*, 2020). This type of adsorption is particularly important for large, nonpolar organic molecules, including many synthetic dyes, pesticides, pharmaceuticals, and volatile organic compounds. These molecules often have planar aromatic structures that can bring them into line with the surface of the adsorbent, enhancing van der Waals interactions (Tang *et al.*, 2018). The low binding strength of physisorption allows easy regeneration of the adsorbent, which is an advantage for a sustainable wastewater treatment system.

5.2.2   Chemical adsorption

Chemical adsorption includes the formation of strong covalent or ionic bonds between the adsorbate and the active sites on the surface of the adsorbent. These interactions are much stronger than van der Waals forces, with adsorption energies typically greater than 80 kJ/mol. As a result, chemisorption is often irreversible or requires sufficient energy for the removal of the molecule (Morrison, 1982). Chemisorption usually occurs on specific active sites and is limited to monolayer coverage because each adsorbate molecule forms a distinct bond. This process often requires activation energy, meaning it may occur more effectively at elevated temperatures. Surface functional groups such as hydroxyl (–OH), carboxyl (–COOH), amine (–NH$_2$), and phosphate groups play an important role in simplifying chemisorption by providing sites for ionic or coordinate bonding (Zhang *et al.*, 2024). This mechanism is dominant for charged or reactive pollutants, especially heavy metals like Cd^{2+}, Cr^{6+}, Pb^{2+}, and As^{3+}. These ions can form inner-sphere complexes with surface ligands, leading to a stable adsorption bond. Chemisorption is essential for applications requiring strong binding, such as the immobilization of toxic metal ions from industrial wastewater.

5.2.3 Electrostatic attraction in adsorption

Electrostatic attraction refers to the force of attraction between oppositely charged particles. In adsorption processes, it plays an important role when adsorbents possess charged surfaces that interact with charged pollutants (ions or ionizable species) present in the solution. This interaction is governed by Coulomb's Law, which states that the force between two charged particles is directly proportional to the product of their charges and inversely proportional to the square of the distance between them (Lekner, 2012). The electrostatic forces are particularly significant in aqueous-phase adsorption systems, where surface charge on the adsorbent and the ionic nature of the pollutant jointly determine the efficiency and direction of adsorption. When an adsorbent surface is positively charged, it attracts negatively charged ions (anions) from the solution. Common anionic pollutants include phosphate (PO_4^{3-}), nitrate (NO_3^-), fluoride (F^-), and anionic dyes such as Congo red or reactive black (Bensalah, 2020). Magnetic biochar and modified adsorbents can be engineered to have positive charges by functionalization with metals like iron, aluminium, or amine groups, enhancing their capacity to remove anions through electrostatic attraction. In contrast, a negatively charged surface draws in positively charged ions (cations) such as Pb^{2+} (lead), Cd^{2+} (cadmium), Cr^{3+} (chromium), Cu^{2+} (copper), or cationic dyes like methylene blue. Many carbon-based adsorbents such as biochar or activated carbon naturally acquire negative surface charges due to the presence of oxygen-containing groups (–OH, –COOH) especially in alkaline media (Atchabarova *et al.*, 2022). This makes them highly effective in capturing metal ions from industrial wastewater.

5.2.4 Influence of solution pH and point of zero charge (pH PZC)

The solution pH and the point of zero charge (pH PZC) of the adsorbent surface are critical factors in determining the surface charge and, hence, the direction and intensity of electrostatic attraction. The pH PZC is defined as the pH at which the net surface charge of an adsorbent is zero (Bakatula *et al.*, 2018). Below this pH, the surface tends to be positively charged, while above it, the surface becomes negatively charged. At pH <pH PZC, the surface is protonated (H^+ ions dominate), promoting the adsorption of anions like phosphate and nitrate due to their opposite charge. At pH >pH PZC, the surface becomes deprotonated (loss of H^+), making it negatively charged and thereby favouring the adsorption of metal cations such as Pb^{2+} or Cd^{2+}. This relationship between pH, surface charge, and pollutant charge enables environmental engineers to tune the adsorption process by adjusting the pH to maximize pollutant removal. Electrostatic attraction is an important mechanism in adsorption, allowing the selective removal of ionic pollutants based on the interplay between surface charge and pollutant charge (Yan *et al.*, 2016).

5.2.5 Ion exchange

Ion exchange is an important adsorption mechanism in which pollutant ions in solution replace pre-existing ions such as Na^+, H^+ on the surface of an adsorbent like biochar. This process is used by active functional groups such as carboxyl (–COOH) and hydroxyl (–OH), which provide negatively charged sites that can attract incoming cations like Pb^{2+} or Cu^{2+}. For example, in aqueous solutions, Pb^{2+} ions can displace Na^+ ions associated with –COO^- groups on the biochar surface, used to effect metal removal (Xu *et al.*, 2022). This exchange does not require pore penetration and often contributes to high selectivity and reversibility in pollutant capture. The occurrence of ion exchange can be confirmed using spectroscopic tools such as Fourier transform infrared (FTIR) spectroscopy, which shows shifts in functional group vibrations as changes in –COO^- peaks, or x-ray photoelectron spectroscopy (XPS), which shows the replacement of surface ions through the arrival of new elemental peaks and changes in prevailing ones.

5.2.6 Surface complexation

Surface complexation is a chemical adsorption mechanism in which pollutant ions are converted to stable functional groups present on the surface of adsorbents such as biochar, activated carbon, or metal oxides. This interaction goes beyond simple physical attraction and involves the sharing or transfer of electrons between the pollutant ion and surface ligands. These functional groups serve as coordination sites where heavy metals or other ionic contaminants bind strongly through covalent or coordinate bonds, forming what is referred to as a surface complex (Vandenbossche *et al.*, 2015). This process is particularly important in the adsorption of heavy metal ions such as Pb^{2+}, Cd^{2+}, Zn^{2+}. These ions have high charge densities and a strong tendency to form chemical bonds with available electron-donating groups. The complexation binds the pollutant to the adsorbent surface, often resulting in irreversible binding – a key advantage for preventing leaching or secondary contamination. Surface complexation can occur in two ways. inner-sphere complexation involves direct coordination between the metal ion and the functional group, with no water molecule in between. This leads to strong, specific binding. Outer-sphere complexation, in which the ion is held through electrostatic interaction, has a layer of water molecules remaining between the surface and the ion, resulting in weaker and more reversible bonding (Fayer, 2012). Overall, surface complexion contributes significantly to the selectivity, strength, and permanence of adsorption, especially in removing toxic metals from wastewater. This mechanism is often supported by FTIR, XPS, or extended X-ray absorption fine structure (EXAFS) analyses, which can reveal changes in bonding and coordination environments after adsorption (Huang *et al.*, 2017).

5.2.7 π–π Interactions

Interactions termed π–π interactions (also called π–π stacking) are a type of non-covalent interaction that occurs between aromatic rings, which are systems of delocalized electrons (π-electrons) found in many organic compounds. These interactions are especially important in adsorption processes involving aromatic pollutants, such as synthetic dyes and pharmaceuticals, because both the pollutant and the adsorbent often contain aromatic structures that can align and interact with each other through their π-electron clouds. In the context of adsorption, π–π interactions occur when the aromatic rings of organic molecules (e.g. dyes such as methylene blue and Congo red) interact with the graphene-like sp^2-hybridized carbon layers present in biochar, activated carbon, or carbon nanotubes. These carbon-based materials contain conjugated π-electron systems that are highly compatible with other aromatic compounds, leading to stacking or overlapping of the electron clouds (Ghasemi and Moth-Poulsen, 2021). This stacking minimizes system energy and promotes a stable non-covalent association between the pollutant and the adsorbent. There are two major types of π–π stacking: face-to-face (parallel) stacking, where the aromatic rings are aligned directly on top of each other; and edge-to-face (T-shaped) stacking, where one ring is perpendicular to another, often seen in mixed systems. These π–π interactions are particularly crucial for the adsorption of aromatic dyes and pharmaceutical residues, because they often govern the affinity and selectivity of the adsorbent material. The effectiveness of this mechanism can be affected by factors such as the number of aromatic rings. Unlike covalent bonding, π–π interactions are reversible and allow for the renewal of the adsorbent, making them valuable in sustainable water treatment applications (Zhao and Zhu, 2020).

5.2.8 Magnetic interaction

Magnetic interaction in adsorption refers to the role of iron-based magnetic particles, such as magnetite (Fe_3O_4) and magnetite (γ-Fe_3O_4), in enhancing pollutant removal through multiple mechanisms. These particles not only allow the easy separation of the adsorbent using an external magnet, but also actively participate in pollutant binding (Feng *et al.*, 2000). They contribute to adsorption via co-precipitation, where metal ions such as Pb^{2+} or arsenate form insoluble complexes with iron species, and through electrostatic interactions, as the charged surfaces of iron oxides attract oppositely charged contaminants. Furthermore, surface functional groups (e.g. Fe–OH) can bind with heavy metals via surface complexion, making magnetic particles highly effective in capturing both cationic and anionic pollutants. The integration of these particles into carbon-based adsorbents results in multifunctional magnetic materials that are not only efficient at pollutant removal but also easily recoverable, making them valuable for sustainable wastewater treatment (Yaqoob and Hanif, 2025).

5.2.9   Redox reactions

Redox reactions (reduction–oxidation reactions) in adsorption systems refer to electron transfer processes that occur between adsorbents and pollutants, particularly when iron oxides (like Fe_3O_4 or γ-Fe_3O_4) are incorporated into the adsorbent. These redox-active iron particles enable catalytic transformations of contaminants, going beyond simple adsorption by altering the chemical form of the pollutant, often into a less toxic or more stable state. Iron oxides contain Fe in multiple oxidation states, primarily Fe^{2+} and Fe^{3+}. This redox flexibility allows them to either donate or accept electrons, making them capable of catalysing redox reactions in environmental systems (Huang *et al.*, 2021). These reactions often occur at the surface of magnetic biochar or iron-modified adsorbents, where pollutants come into contact with redox-active sites. One of the most widely studied redox processes involves the reduction of hexavalent chromium Cr(VI)), a highly toxic and carcinogenic ion, to trivalent chromium (Cr(III)), which is far less toxic and less mobile in the environment. This occurs via electron donation from Fe^{2+} present in magnetite (Fe_3O_4), following the reaction:

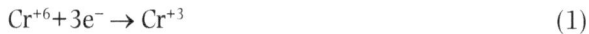

$$Cr^{+6} + 3e^- \rightarrow Cr^{+3} \qquad (1)$$

Iron donates electrons to Cr(VI), converting it into a safer form that often precipitates or becomes adsorbed onto the biochar surface. Iron oxides can also act as oxidants or catalysts in Fenton-like reactions, where hydrogen peroxide (H_2O_2) reacts with Fe^{2+} to generate hydroxyl radicals ($\bullet$OH) – powerful oxidizing agents (Henle *et al.*, 1996). These radicals can degrade persistent organic pollutants such as dyes, phenols, or pharmaceuticals into smaller, less harmful molecules. This mechanism is especially valuable in treating non-biodegradable contaminants that resist conventional biological treatment (Bellanthudawa *et al.*, 2023). A comparison of the different adsorption mechanisms is shown in Table 5.1.

5.3   Mechanisms for Specific Pollutant Types

5.3.1   Heavy metals

Some heavy metals like lead (Pb^{2+}), cadmium (Cd^{2+}), arsenic (As^{3+}), and chromium (Cr^{6+}) are among the most toxic in aquatic environments. Their effective removal using biochar-based adsorbents primarily involves mechanisms like

Table 5.1. Comparison of different adsorption mechanisms.

Adsorption mechanism	Principle	Type	Key applications
Physisorption (physical adsorption)	Involves weak van der Waals forces; no chemical bond formation (Israelachvili, 1974)	Reversible, multilayer	Gas separation, dye removal, activated carbon filters, low-temperature adsorption
Chemisorption (chemical adsorption)	Involves strong chemical bonds (ionic or covalent) with adsorbate (Sun *et al.*, 2017)	Irreversible, monolayer	Catalysis, heavy metal removal, heterogeneous catalysis, corrosion inhibitors
Ion exchange	Exchange of ions between adsorbent surface and solution	Chemical/electrostatic	Water softening, wastewater treatment, removal of cations/anions such as Pb^{2+}, NO_3^-
Electrostatic attraction	Attraction between opposite charges of adsorbent surface and pollutant (Yan *et al.*, 2016)	Physical	Adsorption of dyes, colloids, and charged heavy metals (e.g. Cr^{6+}, methylene blue, Congo red)
Pore filling / surface diffusion	Adsorbate fills micro/mesopores or migrates along the adsorbent surface (Liu *et al.*, 2021)	Physical	Gas storage (H_2, CH_4), volatile organic compound capture, dye molecule entrapment in biochar/activated carbon pores
Hydrogen bonding	Interaction between hydrogen donor/acceptor in adsorbent and adsorbate	Physical/chemical	Organic pollutant adsorption, pharmaceuticals, phenolic compound removal (Kumar *et al.*, 2025)
π–π Interaction	π–Electron interaction between aromatic rings of adsorbent and pollutant	Physical/chemical	Adsorption of aromatic dyes (e.g. methylene blue, rhodamine B), antibiotics, organic molecules

Continued

Table 5.1. Continued

Adsorption mechanism	Principle	Type	Key applications
Ligand exchange	Exchange between ligands on the surface of adsorbent and functional groups of adsorbates	Chemical	Arsenic, phosphate, and fluoride removal from water using metal-oxide-modified adsorbents
Complexation / coordination	Formation of coordination bonds between metal ions and surface groups (Liu *et al.*, 2017)	Chemical	Removal of heavy metals such as Cu^{2+} and Zn^{2+} using functionalized biochars and magnetic adsorbents
Hydrophobic interaction	Non-polar interaction between hydrophobic surfaces and non-polar adsorbate	Physical	Adsorption of oils, organ chlorines, and hydrocarbons from industrial effluents

ion exchange, surface complexation, and electrostatic attraction. Ion exchange occurs when metal ions from the aqueous phase displace ions such as H^+, Na^+, or Ca^{2+} that are loosely bound to the functional groups on the biochar surface. Functional groups like carboxyl (–COOH) and hydroxyl (–OH) serve as active sites where heavy metal cations replace existing surface ions. For instance, Pb^{2+} or Cu^{2+} may displace Na^+ on –COO Na groups, binding more strongly due to higher charge density. Surface complexation is another dominant mechanism where metal ions form inner-sphere or outer-sphere complexes with surface groups. Inner-sphere complexation involves direct bonding between the metal ion and functional group, whereas outer-sphere complexation retains a hydration shell (Lindoy, 1996). Carboxyl, carbonyl, and phosphate groups on biochar often participate in such complexion, creating stable and irreversible binding. Electrostatic attraction also plays a vital role, particularly when the surface of the adsorbent carries a negative charge. At pH values above the point of zero charge (pH PZC), the surface becomes negatively charged and attracts positively charged metal ions. This attraction enhances uptake and improves removal efficiency, especially for cations like Pb^{2+}, Cd^{2+}, and Zn^{2+}. The synergy between these mechanisms enhances the overall adsorption performance and enables the use of magnetic or chemically modified biochar for targeted heavy metal remediation in complex wastewater matrices.

5.3.2 Dyes

Synthetic dyes such as methylene blue, Congo red, and malachite green are large, complex organic molecules widely used in the textile and printing industries. Their removal from wastewater is challenging due to their chemical stability and water solubility. The primary adsorption mechanisms for dye removal using carbon-based adsorbents are π–π stacking, electrostatic interactions, and pore filling. The π–π stacking involves non-covalent interactions between the aromatic rings of the dye molecules and the conjugated π-electron systems of the carbon matrix in the adsorbent, such as in biochar or activated carbon (Jing *et al.*, 2025). These interactions are especially effective for aromatic dyes due to the planarity and electron-rich nature of both the adsorbate and the adsorbent surface. Electrostatic interactions depend on the charge of the dye molecule and the surface charge of the adsorbent, which is pH-dependent. Cationic dyes like methylene blue are adsorbed more efficiently on negatively charged surfaces, whereas anionic dyes like Congo red require positively charged surfaces for strong attraction. The surface charge is influenced by the pH PZC of the adsorbent. Pore filling is a physical process where dye molecules are trapped within the porous network of the adsorbent. High surface area and well-developed porosity facilitate better diffusion and retention of dye molecules within the microspores and mesopores of the material (Lv *et al.*, 2018). This physical entrapment complements chemical interactions, leading to high dye-uptake capacities. The combination of these mechanisms ensures the broad-spectrum adsorption of various dye types, making biochar-based materials suitable for wastewater treatment in the textile industry.

5.3.3 Pharmaceuticals and organic pollutants

Adsorbents like biochar, activated carbon, and carbon nanotubes remove pharmaceutical and organic pollutants primarily through hydrophobic interactions, π–π stacking, and hydrogen bonding (Zaidi *et al.*, 2025). The hydrophobic interactions occur when non-polar segments of pharmaceutical molecules preferentially associate with the hydrophobic carbon surface of the adsorbent. This mechanism is especially effective for poorly soluble organic compounds and enhances the adsorption of large, non-polar drugs. The interaction π–π stacking facilitates the binding of aromatic pharmaceuticals that contain benzene rings with the sp^2 carbon framework of biochar. This interaction allows stable alignment of the π-electron clouds between the adsorbate and adsorbent. Hydrogen bonding occurs between electronegative atoms (such as N or O) in pharmaceutical molecules and hydrogen-donating functional groups (e.g. –OH, –NH_2) on the biochar surface (Aw *et al.*, 2013). These weak bonds enhance specificity and contribute to the selective adsorption of molecules with polar functional groups.

These combined mechanisms provide high selectivity and binding efficiency for a diverse range of pharmaceutical contaminants, particularly when adsorbents are chemically modified to introduce functional groups or increase hydrophobicity (Adewuyi, 2020).

5.3.4 Nutrients (phosphate, nitrate)

Nutrient pollution from excess phosphate and nitrate in agricultural runoff and municipal wastewater leads to eutrophication, algal blooms, and oxygen depletion in aquatic ecosystems. Effective removal of these nutrients is achieved through mechanisms such as precipitation with iron oxides and electrostatic attraction. Phosphate ions have a strong affinity for iron and aluminium oxides commonly embedded in magnetic or metal-modified biochar. These oxides promote the precipitation of phosphate by forming stable complexes such as $FePO_4$ on the adsorbent surface (Cui *et al.*, 2023). The electrostatic attraction also plays an important role, especially under acidic to neutral pH conditions. At pH values below the pH PZC, the adsorbent surface becomes positively charged and can attract negatively charged nitrate (NO_3^-) and phosphate (PO_4^{3-}). The strength of this contact is reliant on the surface charge density and ionic strength of the solution. These mechanisms are often enhanced through chemical modifications or by loading biochar with Fe, Al, or Mg to create composite adsorbents that target nutrient pollutants effectively.

5.4 Factors Affecting the Removal Mechanisms

5.4.1 pH

The pH of the solution is one of the most critical factors affecting adsorption mechanisms, directly affecting the presence of pollutants in solution and the surface charge of the adsorbent. For many contaminants, whether they are ionized or unionized depends on the pH of the surrounding environment. For example, heavy metals may exist as free ions such as Pb^{2+} and Cu^{2+} at lower pH levels but precipitate or form hydroxide complexes at higher pH values. The surface charge of the adsorbent is also highly pH dependent. The pH PZC defines the pH at which the surface has a net neutral charge (Bakatula *et al.*, 2018). Below this point, the surface tends to be positively charged, preferring the adsorption of anions such as nitrate, phosphate, or anionic dyes. Above the pH PZC, the surface becomes negatively charged, which increases the attraction of cationic species like metal ions and positively charged dyes such as methylene blue. Therefore, understanding and altering the pH is essential for maximizing electrostatic attraction and optimizing adsorption efficiency. In addition to electrostatics, pH affects surface complexion and ion exchange as functional groups like –COOH and –OH can undergo protonation or deprotonation. At lower pH, protonation may block active sites, while at higher pH, deprotonation increases the availability of binding sites for metal ions or polar organic molecules. Thus, optimal pH conditions must be selected based on the specific pollutant and adsorbent properties. The effects of process parameters on dye or metal removal by magnetic biochar are shown in Table 5.2.

Table 5.2. Effect of process parameters on dye or metal removal by magnetic biochar.

Process parameter	Effect on adsorption	Mechanism involved	Notes / examples
pH	Affects surface charge of biochar and ionization state of pollutants (Tan et al., 2020)	Electrostatic interaction, ion exchange	Low pH favours cationic dye adsorption (e.g. methylene blue); high pH favours metal ion removal (e.g. Pb^{2+})
Temperature	Can enhance or reduce adsorption depending on whether the process is endothermic or exothermic	Thermodynamic equilibrium (ΔH, ΔG, ΔS)	Higher temperature often increases dye adsorption due to increased diffusion
Initial pollutant concentration	Higher concentration increases driving force for mass transfer but can saturate adsorption sites (Fulazzaky et al., 2013)	Surface diffusion, pore filling	High initial concentrations of dyes like Congo red and methylene blue show saturation trends
Contact time	Determines adsorption rate and equilibrium time	External diffusion, intraparticle diffusion	Rapid uptake initially, then plateau as equilibrium is approached
Adsorbent dose	Higher dose increases available surface area and active sites (Schmid and Stoeger, 2016)	Monolayer/ multilayer coverage	Optimal dose prevents unnecessary overuse of adsorbent
Stirring speed (RPM)	Improves mixing and reduces boundary layer resistance	Mass transfer enhancement	RPM between 200–400 improves dye/metal removal efficiency
Ionic strength / co-existing ions	Can compete with target pollutants for adsorption sites	Competitive adsorption	Presence of Na^+, Ca^{2+}, Cl^- can reduce adsorption efficiency of heavy metals
Magnetization (Fe content)	Affects surface properties and active functional groups (Seredych et al., 2008)	Complexation, redox, electrostatic	Higher Fe^{3+} content improves metal ion adsorption due to coordination

5.4.2 Temperature

Temperature significantly influences adsorption behaviour by affecting both the thermodynamic feasibility and kinetic energy of adsorbate molecules. Adsorption processes can be either exothermic or endothermic and this determines whether an increase in temperature enhances or reduces removal efficiency. In exothermic adsorption, which is common for physical adsorption and electrostatic attraction, increasing temperature may reduce adsorption capacity due to the tendency of the system to favour desorption. In contrast endothermic processes, such as those involving strong chemisorption, surface complexation, or intraparticle diffusion, may benefit from elevated temperatures (Mufazzal Saeed and Ahmed, 2006). Higher thermal energy increases molecular motion, reduces solution viscosity, and improves mass transfer rates, allowing pollutants to reach and interact with adsorbent surfaces more efficiently. Thermodynamic parameters such as enthalpy change (ΔH), Gibbs free energy (ΔG), and entropy change (ΔS) help in understanding the temperature effect. For example, a positive ΔH value indicates an endothermic process, and a negative ΔG signifies spontaneous adsorption. Therefore, selecting appropriate temperature conditions is essential in designing adsorption systems, particularly for industrial wastewater, where heat may be a by-product or cost factor.

5.4.3 Initial pollutant concentration

The initial concentration of pollutants in solution has a direct impact on the driving force for mass transfer and the overall adsorption capacity of the system. At low initial concentrations, the adsorbent surface has a high availability of active sites relative to the number of pollutant molecules, resulting in efficient uptake and high percentage removal. However, as the initial concentration increases, the present active sites become saturated, and the adsorption rate levels off, reflecting a typical Langmuir-type behaviour. A higher concentration also increases the concentration gradient, which serves as the driving force for diffusion, increasing the rate at which pollutants move from the bulk solution to the adsorbent surface. The capacity of the adsorbent is limited and increasing concentration can eventually lead to reduced removal efficiency and the need for more adsorbent or multi-stage treatment systems. In practical applications, understanding how adsorption varies with pollutant concentration is important for sizing equipment, determining operational dosages, and also cost-effectiveness. Adsorption isotherms such as Langmuir, Freundlich, and Temkin models are used to model this concentration dependence and optimize the design of treatment systems (Chung *et al.*, 2015).

5.4.4 Competing ions and co-contaminants

The presence of competing ions and co-contaminants in real wastewater systems introduces complexity to adsorption processes. In natural or industrial wastewater, multiple pollutants are typically present, and these may compete for the same active sites on the adsorbent surface (Raji *et al.*, 2023). For example, in the presence of both Pb^{2+} and Cd^{2+}, their relative affinities and concentrations will influence which ion dominates adsorption. Similarly, cations such as Ca^{2+} and Mg^{2+} in water can compete with target pollutants for ion-exchange sites, reducing overall efficiency. Some ions may inhibit adsorption by occupying or blocking functional groups. For example, phosphate ions may form stable complexes with iron oxides, thereby preventing the adsorption of other anions (Weng *et al.*, 2012). Conversely, certain co-contaminants may enhance adsorption by modifying the adsorbent surface or altering pollutant chemistry. For instance, organic matter can increase hydrophobicity or create new adsorption sites via pre-conditioning.

5.5   Conclusion

The efficiency of pollutant removal through adsorption is governed by several interrelated factors that influence surface interactions, chemical equilibrium, and physical transport processes. Among these, pH plays a dual role by altering both the surface charge of the adsorbent and the chemical form of the pollutant, affecting adsorption selectivity and capacity. As it is well known that temperature control is an important adsorption factor, with adsorption being improved at higher temperatures due to increased molecular mobility and reduced viscosity. Initial pollutant concentration determines the concentration of the mass transfer driving force and the likelihood of surface saturation, while opposing ions and co-contaminants introduce real-world complexity that can either delay or support adsorption. These interactions highlight the importance of optimizing each parameter based on the pollutant type, adsorbent nature, and treatment objectives. The use of modified biochar and composite materials can enhance pathways to improve performance, even in mixed-contaminant systems. There is a need for integrative design approaches and cautious control of environmental conditions to advance adsorption-based water purification technologies.

Acknowledgements

The authors acknowledging their parent institutes at Harcourt Butler Technical University, Kanpur, Uttar Pradesh, India, for providing space, facility and also thank Mr AK for kind cooperation during the manuscript preparation.

References

Adewuyi, A. (2020) Chemically modified biosorbents and their role in the removal of emerging pharmaceutical waste in the water system. *Water* 12(6), 1551. DOI: 10.3390/w12061551.

Akhtar, M.S., Ali, S. and Zaman, W. (2024) Innovative adsorbents for pollutant removal: Exploring the latest research and applications. *Molecules (Basel, Switzerland)* 29(18), 4317. DOI: 10.3390/molecules29184317.

Alaqarbeh, M. (2021) Adsorption phenomena: Definition, mechanisms, and adsorption types: Short review. *RHAZES: Green and Applied Chemistry* 13, 43–51. DOI: 10.48419/IMIST.PRSM/rhazes-v13.28283.

Atchabarova, A.A., Abdimomyn, S.K., Abduakhytova, D.A., Zhigalenok, Y.R., Tokpayev, R.R. *et al.* (2022) Role of carbon material surface functional groups on their interactions with aqueous solutions. *Journal of Electroanalytical Chemistry* 922, 116707. DOI: 10.1016/j.jelechem.2022.116707.

Aw, M.S., Bariana, M., Yu, Y., Addai-Mensah, J. and Losic, D. (2013) Surface-functionalized diatom microcapsules for drug delivery of water-insoluble drugs. *Journal of Biomaterials Applications* 28(2), 163–174. DOI: 10.1177/0885328212441846.

Bakatula, E.N., Richard, D., Neculita, C.M. and Zagury, G.J. (2018) Determination of point of zero charge of natural organic materials. *Environmental Science and Pollution Research International* 25(8), 7823–7833. DOI: 10.1007/s11356-017-1115-7.

Bellanthudawa, B.K.A., Nawalage, N.M.S.K., Handapangoda, H.M.A.K., Suvendran, S., Wijayasenarathne, K.A.S.H. *et al.* (2023) A perspective on biodegradable and non-biodegradable nanoparticles in industrial sectors: Applications, challenges, and future prospects. *Nanotechnology for Environmental Engineering* 8(4), 975–1013. DOI: 10.1007/s41204-023-00344-7.

Bensalah, H. (2020) Natural and synthetic apatites as adsorbents for the removal of azo dyes from aqueous solutions. Doctoral dissertation, Technische Universität, Berlin, Germany.

Björneholm, O., Öhrwall, G., De Brito, A.N., Ågren, H. and Carravetta, V. (2022) Superficial tale of two functional groups: On the surface propensity of aqueous carboxylic acids, alkyl amines, and amino acids. *Accounts of Chemical Research* 55(23), 3285–3293. DOI: 10.1021/acs.accounts.2c00494.

Chung, H.K., Kim, W.H., Park, J., Cho, J., Jeong, T.Y. *et al.* (2015) Application of Langmuir and Freundlich isotherms to predict adsorbate removal efficiency or required amount of adsorbent. *Journal of Industrial and Engineering Chemistry* 28, 241–246. DOI: 10.1016/j.jiec.2015.02.021.

Compaan, K., Vergenz, R., Von Rague Schleyer, P. and Arreguin, I. (2008) Carbon-donated hydrogen bonding: Electrostatics, frequency shifts, directionality, and bifurcation. *International Journal of Quantum Chemistry* 108(15), 2914–2923. DOI: 10.1002/qua.21811.

Cui, J., Yang, J., Weber, M., Yan, J., Li, R. *et al.* (2023) Phosphate interactions with iron-titanium oxide composites: Implications for phosphorus removal/recovery from wastewater. *Water Research* 234, 119804. DOI: 10.1016/j.watres.2023.119804.

Dutta, S., Fajal, S. and Ghosh, S.K. (2024) Heavy metal-based toxic oxo-pollutants sequestration by advanced functional porous materials for safe drinking water. *Accounts of Chemical Research* 57(17), 2546–2560. DOI: 10.1021/acs.accounts.4c00348.

Fayer, M.D. (2012) Dynamics of water interacting with interfaces, molecules, and ions. *Accounts of Chemical Research* 45(1), 3–14. DOI: 10.1021/ar2000088.

Feng, D., Aldrich, C. and Tan, H. (2000) Removal of heavy metal ions by carrier magnetic separation of adsorptive particulates. *Hydrometallurgy* 56(3), 359–368. DOI: 10.1016/S0304-386X(00)00085-2.

Fulazzaky, M.A., Khamidun, M.H. and Omar, R. (2013) Understanding of mass transfer resistance for the adsorption of solute onto porous material from the modified mass transfer factor models. *Chemical Engineering Journal* 228, 1023–1029. DOI: 10.1016/j.cej.2013.05.100.

Ghasemi, S. and Moth-Poulsen, K. (2021) Single molecule electronic devices with carbon-based materials: Status and opportunity. *Nanoscale* 13(2), 659–671. DOI: 10.1039/d0nr07844a.

Henle, E.S., Luo, Y. and Linn, S. (1996) Fe^{2+}, Fe^{3+}, and oxygen react with DNA-derived radicals formed during iron-mediated Fenton reactions. *Biochemistry* 35(37), 12212–12219. DOI: 10.1021/bi961235j.

Huang, J., Jones, A., Waite, T.D., Chen, Y., Huang, X. *et al.* (2021) Fe (II) redox chemistry in the environment. *Chemical Reviews* 121(13), 8161–8233. DOI: 10.1021/acs.chemrev.0c01286.

Huang, Z., Li, Z., Zheng, L., Zhou, L., Chai, Z. *et al.* (2017) Interaction mechanism of uranium (VI) with three-dimensional graphene oxide-chitosan composite: Insights from batch experiments, IR, XPS, and EXAFS spectroscopy. *Chemical Engineering Journal* 328, 1066–1074. DOI: 10.1016/j.cej.2017.07.067.

Israelachvili, J.N. (1974) The nature of van der Waals forces. *Contemporary Physics* 15(2), 159–178. DOI: 10.1080/00107517408210785.

Jing, L., Li, P., Li, Z., Ma, D. and Hu, J. (2025) Influence of π-π interactions on organic photocatalytic materials and their performance. *Chemical Society Reviews* 54(4), 2054–2090. DOI: 10.1039/d4cs00029c.

Kumar, A., Kapoor, A., Kumar Rathoure, A., Devnani, G.L. and Pal, D.B. (2025) Enhanced surface properties of biochar using activation strategies for sustainable dye removal: A review. *Asia-Pacific Journal of Chemical Engineering* e70122. DOI: 10.1002/apj.7012.

Lekner, J. (2012) Electrostatics of two charged conducting spheres. *Proceedings of the Royal Society A: Mathematical, Physical and Engineering Sciences* 468(2145), 2829–2848. DOI: 10.1098/rspa.2012.0133.

Lindoy, L.F. (1996) Outer-sphere and inner-sphere complexation of cations by the natural ionophore lasalocid A. *Coordination Chemistry Reviews* 148, 349–368. DOI: 10.1016/0010-8545(95)01192-7.

Liu, P., Qin, R., Fu, G. and Zheng, N. (2017) Surface coordination chemistry of metal nanomaterials. *Journal of the American Chemical Society* 139(6), 2122–2131. DOI: 10.1021/jacs.6b10978.

Liu, Y., Qi, R., Ge, Z., Zhang, Y., Jing, L. *et al.* (2021) N-doping copolymer derived hierarchical micro/mesoporous carbon: Pore regulation of melamine and fabulous adsorption performances. *Journal of the Taiwan Institute of Chemical Engineers* 120, 236–245. DOI: 10.1016/j.jtice.2021.03.030.

Lu, L., Yu, W., Wang, Y., Zhang, K., Zhu, X. *et al.* (2020) Application of biochar-based materials in environmental remediation: From multi-level structures to specific devices. *Biochar* 2(1), 1–31. DOI: 10.1007/s42773-020-00041-7.

Lv, L., Huang, Y. and Cao, D. (2018) Nitrogen-doped porous carbons with ultrahigh specific surface area as bifunctional materials for dye removal of wastewater and supercapacitors. *Applied Surface Science* 456, 184–194. DOI: 10.1016/j. apsusc.2018.06.116.

Morrison, S.R. (1982) Chemisorption on nonmetallic surfaces. In: *Catalysis: Science and Technology.* Springer, Berlin, Heidelberg, pp. 199–229. DOI: 10.1007/978-3-642-93223-6_4.

Mufazzal Saeed, M. and Ahmed, M. (2006) Effect of temperature on kinetics and adsorption profile of endothermic chemisorption process: –tm(iii)–pan loaded puf system. *Separation Science and Technology* 41(4), 705–722. DOI: 10.1080/01496390500527993.

Nagal, B. and Prabhakar, A.K. (2025) Sustainable water management in the digital age: Intelligent solutions for a precious resource. In: *Intelligent Infrastructure and Smart Materials: Sustainable Technologies for a Greener Future.* Springer Nature Switzerland, Cham, pp. 91–114.

Raji, Z., Karim, A., Karam, A. and Khalloufi, S. (2023) Adsorption of heavy metals: Mechanisms, kinetics, and applications of various adsorbents in wastewater remediation—a review. *Waste* 1(3), 775–805. DOI: 10.3390/waste1030046.

Schmid, O. and Stoeger, T. (2016) Surface area is the biologically most effective dose metric for acute nanoparticle toxicity in the lung. *Journal of Aerosol Science* 99, 133–143. DOI: 10.1016/j.jaerosci.2015.12.006.

Seredych, M., Hulicova-Jurcakova, D., Lu, G.Q. and Bandosz, T.J. (2008) Surface functional groups of carbons and the effects of their chemical character, density and accessibility to ions on electrochemical performance. *Carbon* 46(11), 1475–1488. DOI: 10.1016/j.carbon.2008.06.027.

Shaheen, S.M., Mosa, A., Natasha, Abdelrahman, H., Niazi, N.K. *et al.* (2022) Removal of toxic elements from aqueous environments using nano zero-valent iron- and iron oxide-modified biochar: A review. *Biochar* 4(1), 24. DOI: 10.1007/s42773-022-00149-y.

Sun, Z., Chai, L., Shu, Y., Li, Q., Liu, M. *et al.* (2017) Chemical bond between chloride ions and surface carboxyl groups on activated carbon. *Colloids and Surfaces A: Physicochemical and Engineering Aspects* 530, 53–59. DOI: 10.1016/j. colsurfa.2017.06.077.

Takio, N., Basumatary, D., Yadav, M. and Yadav, H.S. (2021) Electrostatic and van der waals forces. In: *Biochemistry: Fundamentals and Bioenergetics.* Bentham Science Publishers, pp. 55–89.

Tan, Z., Yuan, S., Hong, M., Zhang, L. and Huang, Q. (2020) Mechanism of negative surface charge formation on biochar and its effect on the fixation of soil Cd. *Journal of Hazardous Materials* 384, 121370. DOI: 10.1016/j.jhazmat.2019.121370.

Tang, H., Zhao, Y., Shan, S., Yang, X., Liu, D. *et al.* (2018) Theoretical insight into the adsorption of aromatic compounds on graphene oxide. *Environmental Science* 5(10), 2357–2367. DOI: 10.1039/C8EN00384J.

Vandenbossche, M., Jimenez, M., Casetta, M. and Traisnel, M. (2015) Remediation of heavy metals by biomolecules: A review. *Critical Reviews in Environmental Science and Technology* 45(15), 1644–1704. DOI: 10.1080/10643389.2014.966425.

Wang, X., Cook, R., Tao, S. and Xing, B. (2007) Sorption of organic contaminants by biopolymers: Role of polarity, structure and domain spatial arrangement. *Chemosphere* 66(8), 1476–1484. DOI: 10.1016/j.chemosphere.2006.09.004.

Weng, L., Van Riemsdijk, W.H. and Hiemstra, T. (2012) Factors controlling phosphate interaction with iron oxides. *Journal of Environmental Quality* 41(3), 628–635. DOI: 10.2134/jeq2011.0250.

Xu, Y., Qu, Y., Yang, Y., Qu, B., Shan, R. *et al.* (2022) Study on efficient adsorption mechanism of Pb^{2+} by magnetic coconut biochar. *International Journal of Molecular Sciences* 23(22), 14053. DOI: 10.3390/ijms232214053.

Yan, H., Yang, H., Li, A. and Cheng, R. (2016) pH-tunable surface charge of chitosan/graphene oxide composite adsorbent for efficient removal of multiple pollutants from water. *Chemical Engineering Journal* 284, 1397–1405. DOI: 10.1016/j.cej.2015.06.030.

Yaqoob, S. and Hanif, M.A. (2025) Use of magnetic nanomaterials in wastewater treatment: A review. *Environmental Monitoring and Assessment* 197(8), 917. DOI: 10.1007/s10661-025-14292-z.

Zaidi, N., Mir, M.A., Chang, S.K., Abdelli, N., Hasnain, S.M. *et al.* (2025) Pharmaceuticals and personal care products as emerging contaminants: Environmental fate, detection, and mitigation strategies. *International Journal of Environmental Analytical Chemistry* 1–29. DOI: 10.1080/03067319.2025.2484456.

Zhang, X., Xiong, Y., Wang, X., Wen, Z., Xu, X. *et al.* (2024) MgO-modified biochar by modifying hydroxyl and amino groups for selective phosphate removal: Insight into phosphate selectivity adsorption mechanism through experimental and theoretical. *Science of The Total Environment* 918, 170571. DOI: 10.1016/j.scitotenv.2024.170571.

Zhao, G. and Zhu, H. (2020) Cation-π interactions in graphene-containing systems for water treatment and beyond. *Advanced Materials* 32(22), e1905756. DOI: 10.1002/adma.201905756.

www.ingramcontent.com/pod-product-compliance
Lightning Source LLC
Chambersburg PA
CBHW042315210326
41599CB00038B/7142